Adaptive 3D Sound Systems

John Garas

Distributors for North, Central and South America:
Kluwer Academic Publishers
101 Philip Drive
Assinippi Park
Norwell, Massachusetts 02061 USA
Telephone (781) 871-6600
Fax (781) 871-6528
E-Mail <kluwer@wkap.com>

Distributors for all other countries:
Kluwer Academic Publishers Group
Distribution Centre
Post Office Box 322
3300 AH Dordrecht, THE NETHERLANDS
Telephone 31 78 6392 392
Fax 31 78 6546 474
E-Mail services@wkap.nl>

Electronic Services <http://www.wkap.nl>

Library of Congress Cataloging-in-Publication

Garas, John, 1966-
Adaptive 3D sound systems / John Garas.
p. cm. -- (Kluwer international series in engineering and computer science ; SECS 566)
Includes bibliographical references and index.
ISBN 0-7923-7907-1
1. Surround-sound systems. 2. Sound--Recording and reproducing. 3. Adaptive filters.
4. Algorithms. I. Title. II. Series.

TK7881.83. G35 2000
621.389'3--dc21

00-041594

Printed on acid-free paper.

Printed in the United States of America

Preface

Stereo sound systems are able to position the perceived sound image at any point on the line between the two reproduction loudspeakers. On the contrary, three-dimensional (3D) sound systems are designed to position the sound image at any point in a 3D listening space around the listener. Processing a monophonic sound signal by filters containing directional cues moves the perceived sound image to the position defined by the filters. The perceived sound source at this position is, therefore, referred to as a virtual or phantom sound source. It is the main objective of this book to investigate the application of adaptive filters in creating robust virtual sound images in real-time through loudspeakers. Adaptive filters are attractive in such applications for two main reasons. The first is their tracking capability. Adaptive filters may be used to track moving listeners and adjust their coefficients accordingly. The second is allowing *in-situ* design of the filters, which in turn allows including the listeners' own head-related transfer functions in the filters design.

To the time of this writing, most work in 3D sound systems has been performed based on anechoic listening situations, which simplifies the analysis considerably. However, this often leads to systems that function properly only in anechoic environments. To avoid such conditional results in the present work, listening spaces are assumed to be reverberant rooms. The acoustic impulse responses measured in those rooms are understood to have long duration due to reverberation. Although this may frustrate the real-time implementation, it permits studying the effects of the listening environment on the system performance. It also motivates searching for the most efficient implementation of the system filters.

Stereo and conventional 3D sound systems are often designed to properly operate for a single listener who is assumed to be seated at an ideal position in the listening space. In the present work, adaptive 3D sound systems are introduced in the general context of multichannel sound reproduction. No assumptions are made concerning the number of listeners, nor their positions in the reverberant listening space. Although this choice complicates the analysis, it results in a general framework that is valid for many applications in the general listening situation of multiple listeners listening to multiple audio signals through multiple loudspeakers.

Since the main goal of 3D sound systems is to convince their users (listeners) that the sound is emitted from a position in the listening space where no physical loudspeaker exists, not only physical parameters but also human perception must be taken into account in designing such system to successfully achieve this goal. Since the output of a 3D sound system is the perceived sound image location, system performance must be tested based on well-designed psychoacoustical experiments. Although simulations have been used throughout this work to show performance improvement, psychoacoustical experiments are inevitable in proving the significance of the methods developed. Those experiments have not been performed in the present work. However, I hope that psychoacousticians will find those methods worth the experiments in the near future. In the next few paragraphs, the contributions of this work to the field of adaptive 3D sound systems are summarised.

Modern 3D sound systems achieve their objectives by filtering a set of monophonic audio signals by a matrix of digital filters. The solution to those filters is, in general, composed of two components. The first is the cross-talk cancellation network, which inverts the matrix of acoustic transfer functions between the reproduction loudspeakers and the listeners' ears. The second is the matrix of desired transfer functions, which contains the directional information the listeners should perceive. Conventional 3D sound systems make use of fixed filters to implement both components. The coefficients of the filters are derived from pre-measured transfer functions, often called the Head-Related Transfer Functions (HRTFs). HRTFs vary significantly among people due to the differences in size and geometrical fine structure of human outer ears. Since it is almost impossible to measure and store the HRTFs for each listener, the fixed filters are usually calculated using generalised HRTFs. These generalised HRTFs are usually obtained by measuring the HRTFs of an artificial head or by averaging the HRTFs measured for several human subjects. Unless the generalised HRTFs match those of the listener, calculating the filters using generalised HRTFs may degrade the perceived sound image.

The above mentioned fixed filters are valid only for a single listening position, namely, the position at which the HRTFs used to calculate the filters have been previously measured. If the listener moves slightly from this ideal position, the filters become in error. Using head trackers, conventional 3D sound systems are able to switch between different pre-designed filters to correct the errors introduced by listener's movements. This requires measuring and storing HRTFs from several discrete directions around the listener.

An alternative approach to 3D sound systems is to replace the fixed filters by adaptive ones. This does not require the above mentioned measurements and operates exclusively using the listeners' own HRTFs. By adjusting the filters' coefficients during system operation, errors introduced due to listeners' movements are automatically corrected. For successful adaptation, a measure of the error criteria must be available. This measure may be obtained by inserting probe microphones inside (or near the entrance of) the listeners' ear canals.

In reverberant environments, the impulse response between two points may last for several hundreds of milliseconds. At 44.1 kHz sampling frequency (the standard for audio compact discs), thousands of FIR filter coefficients are needed to properly model such an impulse response. Therefore, adaptive 3D sound systems require a huge amount of computations, which makes real-time implementation on reasonable hardware resources a difficult task. Reducing the system complexity is, therefore, essential. Implementing the filtering and adaptation operations in the frequency domain reduces the computational complexity considerably. In multichannel systems, further computational saving may be achieved by replacing the conventional filtered-x LMS algorithm by its efficient counterpart, the adjoint LMS algorithm.

For successful tracking of listeners' movements by on-line adaptation of the adaptive filters, the convergence speed of the adaptive algorithm needs to be faster than the listeners' movements. Both filtered-x and adjoint LMS algorithms are known to be slow due to the coloration introduced by filtering through the HRTFs. Convergence speed of those algorithms may be improved by reducing or totally removing this coloration.

System robustness to small listeners' movements may be improved by enlarging the zones of equalisation created around the listeners' ears. Increasing the number of reproduction loudspeakers and correctly positioning those loudspeakers in the listening space improve the system robustness considerably. The filters' behaviour may also be controlled by applying spatial derivative constraints. Alternatively, filters having decreasing frequency resolution with increasing frequency may be employed for the same purpose. Replacing the constant resolution filters with multiresolution ones is also motivated by the fact that the human auditory system performs a similar multiresolution spectral analysis on sound signals.

Multiresolution adaptive filters may be obtained if the filters are implemented in a transformation domain which has the required frequency resolution. To maintain the low computational complexity achieved by implementing the filters in the frequency domain, the multiresolution transformation must be calculated as fast as the Fast Fourier Transformation (FFT). An efficient implementation of the unitary warping transformation using non-uniform sampling is developed for this purpose. By non-uniformly sampling a segment of a signal and calculating the conventional FFT of those samples, a multiresolution transform of that segment is obtained.

Contents

Chapter 1

Introduction

Typical applications of 3D sound systems complement, modify, or replace sound attributes that occur in natural to gain control on one's spatial auditory perception. This control can not be achieved based on physical attributes only. Psychoacoustic considerations of human sound localisation also play important roles in analysing, designing, and testing 3D sound systems. To manipulate one's spatial auditory perception, a thorough understanding of the psychoacoustical phenomena occurring in natural spatial hearing is essential. By modifying the physical parameters associated with those phenomena, the control goal may be achieved. This chapter introduces the problem of designing 3D sound systems, the challenges faced, and the approach followed in the present work. Natural and virtual spatial hearing and the synthesis of localisation cues using conventional digital filters are introduced in Section 1.1. Implementation using adaptive filters, the challenges and advantages in using such filters are mentioned in Section 1.2. In the present work, special attention has been paid to studying the robustness problem, which is inherent to all 3D sound systems. The notion of this problem and the approach followed to contain it are mentioned in Section 1.3. In Section 1.4, example applications of 3D sound systems are considered. Finally, the conventions used throughout the text are listed in Section 1.5.

1.1 Analysis and Synthesis of Spatial Hearing

1.1.1 Natural Spatial Hearing

In natural listening situations, human listeners are capable of estimating the position of the objects radiating the sound waves. The precision of estimating the position of sound sources located in the horizontal plane containing the listener's ears is fairly accurate to allow relying on spatial hearing in everyday life. A typical example often used to explain this sound localisation ability is when one crosses the street looking to the right. Correctly estimating the position and speed of a car coming from the left, from its sound only, helps avoiding an accident.

Using psychoacoustic experiments [30], the important cues used by the human auditory system to localise sound sources in space are identified. It is widely believed that localisation of sources in the horizontal plane containing the listener's ears is due to the differences between the sound waves received at each ear. Localisation of sound sources not in the horizontal plane is strongly influenced by the acoustic filtering due to reflection of short wavelength sound waves from the listener's outer ears and shoulders. Those and other localisation cues are discussed briefly in Chapter 2. The effect of room reverberation on natural spatial listening is also addressed in Chapter 2.

1.1.2 Virtual Spatial Hearing

All *physical* parameters associated with the localisation cues that occur in natural listening situations are embedded in the pair of acoustic transmissions from the source to each of the listener's eardrums. Assume that those acoustic transmissions, usually referred to as the Head-Related Transfer Function (HRTF) pair, can be accurately measured and stored in the form of digital filters. Filtering a monophonic sound signal that contains no directional information through the previously measured HRTF pair results in a pair of electric signals that contain all the information concerning the source location. Playing this pair of electric signals (e.g. through headphones) at the listener's eardrums reproduces the localisation cues, which makes the listener perceive the sound to be coming from the position where the source was at the measurement time. This forms the basic principle of modern 3D sound systems

[25, 54, 72, 143]. The perceived sound source in this listening situation is referred to as a virtual (or phantom) source. The physical properties of HRTF pairs and the synthesis of virtual sound sources are discussed in more details in Chapter 2.

1.1.3 Conventional 3D Sound Systems

To complete the control task, the pair of electric signals containing the directional information must be transformed to acoustic waves prior playing them at the listener's eardrums. This may be accomplished using headphones or a pair of loudspeakers. Headphones allow delivering the electric signals at the listener's ears with minimum cross-talk. Although this simplifies the reproduction process considerably, loudspeakers are preferred in most applications. When loudspeakers are used, the sound from each loudspeaker is heard by both ears. The 3D sound system must, therefore, employ a mechanism to eliminate or reduce this cross-talk [50, 79, 104]. This is usually accomplished by employing a second set of filters often called the cross-talk canceller. Therefore, a 3D sound system that delivers its output sound through loudspeakers employs control filters that are the combination of the cross-talk canceller filters and the HRTF pair containing the directional information. The basic principles of designing the filters employed in conventional (opposed to adaptive) 3D sound systems are discussed in Section 2.2.

1.2 Adaptive 3D Sound Systems

Although several types of fixed digital filters have been used successfully in 3D sound systems [72, 84, 85, 86], adaptive filters may be preferred in many applications. Unlike conventional fixed filters, adaptive filters allow *in-situ* calculation of the control filters. This enables recalculating the filters in response to changes in the listening environment by simply readapting the filters. In Chapter 3, adaptive 3D sound systems are discussed in detail in the context of multichannel sound reproduction.

The adaptation of the system filters in multichannel 3D sound systems is complicated by the acoustic transmissions between the loudspeakers and the listeners' ears. Due to these acoustic transmissions, the conventional Least Mean Square (LMS) algorithm becomes unstable and can not be used. A related algorithm, the filtered-x LMS algorithm, has to be employed. The convergence properties of this algorithm are strongly influenced by the physical properties of the acoustic transmissions. Furthermore, adaptive multichannel 3D sound systems require a huge amount of computations to perform real-time filtering and adaptation operations. Efficient implementation of those operations is, therefore, essential. Reducing the computational complexity of multichannel systems may be achieved by using a more efficient version of the filtered-x LMS algorithm, the adjoint LMS. Further computational saving may be gained by performing the filtering and adaptation operations in the frequency domain. The optimum LMS solution to the control filters, the effects of the acoustic transmissions, the filtered-x LMS, the adjoint LMS, and the frequency domain implementations are treated in greater detail in Chapter 3.

1.3 Robustness of 3D Sound Systems

Since the basic principle of 3D sound systems mentioned above relies on the directional information embedded in the HRTF pairs, it is obvious that the control filters depend on the position of the sound source relative to the listener, independent of the method used to calculate those filters. A set of filters designed for optimum results at a specific listener position becomes in error when the listener moves. This problem, referred to as the robustness problem hereafter, is inherent to all 3D sound systems. The robustness problem and its remedies are investigated in Chapter 4.

Small listener movements are contained by designing the control filters to be valid not only at the eardrums of the listener, but also at points in space in the vicinity of the listener's ears. This is achieved by employing spatial derivative constraints, which limit the sound field to be the same at and around the listener's ears. The effect of these spatial derivative constraints is to force the filters to assume increasingly deviating solutions from the optimum solutions as the frequency increases. Implementations of spatial derivative constraints require measuring several transfer functions in the vicinity of the ears and calculating the control filters using a weighted average of those measurements.

A more efficient approach to forcing the filters to the above mentioned solutions that increasingly deviate from the optimum solutions with increasing frequency may be achieved by using control filters of non-uniform frequency resolution. Filters having coarser frequency resolution with increasing frequency increasingly ignore details in the acoustic transmissions with increasing frequency. This is also in full agreement with the multiresolution analysis performed by the human ears [148]. Multiresolution spectral analysis of a segment of a signal may be calculated by sampling the signal at non-uniformly spaced time moments and calculating the FFT of those samples. This non-uniform sampling method and other techniques for designing filters having such non-uniform frequency resolution are discussed in Chapter 5.

The performance of a sound reproduction system is strongly influenced by the number of reproduction loudspeakers and the positions of those loudspeakers in the listening space. Using computer simulations, it is shown in Chapter 4 that better performance can be obtained by increasing the number of reproduction loudspeakers. Examination of the optimum positions of the loudspeakers is also considered in anechoic and reverberant listening environments.

When a listener makes a large movement, the control filters become in great errors, and new filters that are derived from the new listener's position must be used. In adaptive 3D sound systems, the required filters in the new position may be calculated by readapting the filters to their new optimal solutions. This requires measuring the acoustic transmissions between the loudspeakers and the listener's eardrums. For successful operation, the adaptive filters must be updated faster than the listener's movements. Therefore, improving the convergence speed of the adaptive algorithm is essential. On-line estimation of acoustic transmissions, on-line adaptation of the control filters, and improving convergence speed of the adaptive algorithm are treated at length in Chapter 4.

1.4 Applications

3D sound systems are important in many applications such as consumer electronics, interactive entertainment, desktop computers, multimedia, and human-machine interfaces. In this section, examples of those applications are mentioned.

A simple application of 3D sound systems is the stereo-base widening [1]. The physical dimensions of a television set or a computer monitor may limit the distance between the built-in stereo loudspeakers. The quality of the stereo effect is known to degrade as the distance between the two loudspeakers decreases. Virtual sound sources may be used to achieve a virtual increase in the distance between the two loudspeakers.

Virtual sound sources may also be used for surround sound reproduction using fewer physical loudspeakers than normally required. Current surround sound systems require four or five loudspeakers. The DVD (Digital Versatile Disk) medium allows storing eight audio tracks, which requires eight loudspeakers for sound reproduction. In most home listening environments, this eight-loudspeaker acoustical arrangement may not be convenient. In such situations, virtual sound sources can be generated to replace physical loudspeakers. Since two real loudspeakers can be used to generate N virtual sound sources (see Chapter 2), surround sound can be generated using two front loudspeakers only. This leads to an efficient use of the hardware in applications such as televisions, home theatre, and audio equipment.

Virtual sources are also essential in game machines and virtual reality systems. The experience of a car race game is closer to reality when sounds travel with the cars compared to the case where sounds always come from the same fixed place of the system loudspeaker.

An example of improving human-machine interface using 3D sound systems may be found in the presentation of signals a pilot may receive in his cockpit [25]. Virtual sources may be used to let the pilot perceive the sound of different signals to be coming from different directions. Therefore, decreasing the probability of misunderstanding the messages.

1.5 Conventions

Throughout the text, lower-case letters are used to represent time domain signals while upper-case letters represent frequency domain signals. Boldface characters (eg. $\mathbf{x}$) are used to represent matrixes and underlined boldface characters (eg. $\underline{\mathbf{x}}$) represent vectors. In single channel contexts, matrix and vector notations are used to represent samples of a signal's time history or samples of the Fourier transform of the signal.

For example, the last B samples from the time history of the signal $x_i(n)$ is represented by

$$\underline{\mathbf{x}}_i(n) = \left[\begin{array}{cccccc} x_i(n) & x_i(n-1) & \cdots & x_i(n-b) & \cdots & x_i(n-B+1) \end{array}\right]^T, \tag{1.1}$$

where $\cdot^T$ denotes transpose. Calculating the B-point Discrete Fourier Transformation (DFT) of $\underline{\mathbf{x}}_i(n)$ results in the vector of frequency samples $\underline{\mathbf{X}}_i(\omega)$

$$\underline{\mathbf{X}}_i(\omega) = \left[\begin{array}{cccccc} X_i(\omega_0) & X_i(\omega_1) & \cdots & X_i(\omega_b) & \cdots & X_i(\omega_{B-1}) \end{array}\right]^T, \tag{1.2}$$

where $\omega_b = \frac{2\pi b}{B}$, $b = 0, 1, 2, \cdots, B-1$. In multichannel contexts, matrix and vector notations will be used to represent multiple signals at a single time moment or a single frequency. For example, the set of signals $\{x_i(n) \ : \ i = 1, 2, \cdots, I\}$ at a time sample n is represented by the vector

$$\underline{\mathbf{x}}(n) = \left[\begin{array}{cccc} x_1(n) & x_2(n) & \cdots & x_I(n) \end{array}\right]^T. \tag{1.3}$$

The time history of each signal $x_i(n)$ is expressed by (1.1). Calculating the B-point DFT of the time history of each of the $x_i(n)$ signals and evaluating the result at one specific frequency ω results in the frequency vector $\underline{\mathbf{X}}(\omega)$

$$\underline{\mathbf{X}}(\omega) = \left[\begin{array}{cccc} X_1(\omega) & X_2(\omega) & \cdots & X_I(\omega) \end{array}\right]^T. \tag{1.4}$$

In some cases as in Chapter 3, the time histories of multiple signals are needed to be processed simultaneously. In these cases, composite vectors and matrixes that stack time samples of all signals in specific structures will be defined.

Many concepts are readily understood using discrete-time representations while others are best explained using continuous-time representations of signals. While the discrete-time representations are considered to be the default throughout the text, the continuous-time representations are used whenever necessary. This will be often the case in Chapter 5, which deals with non-uniform sampling and multiresolution spectral analysis.

Frequent use will be made of the Fourier transformation. Therefore, the Fourier operator will be denoted by the special symbol $\mathbb{F}$. In the discrete-time domain, $\mathbb{F}$ is a square matrix which has its (n, m) element defined by

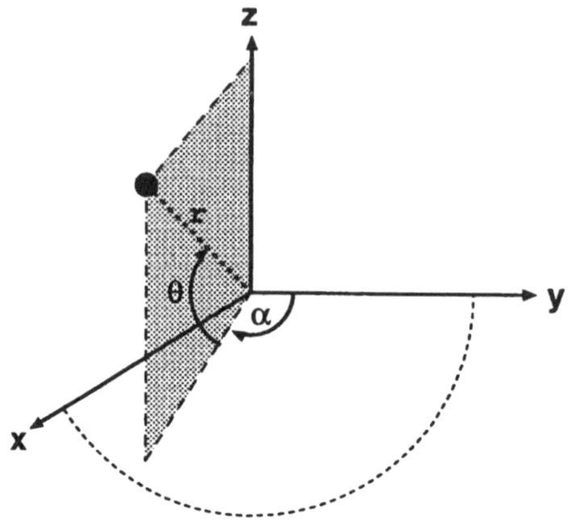

Figure 1.1: The vertical-polar coordinate system.

$e^{-j\frac{2\pi}{B}nm}$, where $j = \sqrt{-1}$ and B is the transformation length. The same symbol will be used for the Fourier transformation in the continuous-time domain and will be defined when needed. Other operators that will be frequently used are the convolution operator denoted by $*$ and the complex conjugate operator denoted by $\cdot^*$.

Both the Cartesian and vertical-polar coordinates will be used throughout the text to express the position of objects in the listening space as shown in Fig. 1.1. In single listener situations, the coordinate system is chosen so that the origin is at the centre point between the listener's ears (the interaural axis). The $+x$-axis passes through the listener's right ear, while the $+y$-axis passes through the listener's nose. The xy, yz, and xz planes are referred to as the horizontal, median, and frontal planes, respectively. In the vertical-polar coordinate system, the azimuth angle α is measured from the median plane to a vertical plane containing the z-axis and the object as shown in Fig. 1.1. The elevation angle θ is then the angle from the horizontal plane to the object on that plane.

Chapter 2

3D Sound Reproduction

Psychoacousticians distinguish between a *sound source* location and an *auditory event* location. The former is the position of a physical sound source in the listening space, while the latter is the position where a listener experiences the sound [30]. From everyday experience we know that a monophonic audio signal played through a loudspeaker makes the sound source and the auditory event positions coincide. However, it is possible to process the audio signal so that the auditory event occurs at a completely different position in the listening space than the position of the physical loudspeaker emitting the sound. The listener perceives the sound to be coming from the auditory event position, which is, therefore, referred to as a phantom (or virtual) sound source. A simple form of this audio processing is the stereophonic audio system [6], where the amplitude or phase of the sound is panned between two loudspeakers. Stereophonic systems are able to position the auditory event (sound image) at any point on the line connecting the two loudspeakers. A direct extension to this technique is the surround sound system, where more than two loudspeakers surrounding the listener are used. By panning the sound between every two adjacent loudspeakers, the auditory event can be positioned on lines connecting the loudspeakers [121]. As the number of reproduction loudspeakers increases, the auditory event can be accurately placed at any point in a three-dimensional (3D) space. This is exploited in the wave field synthesis technique that is based on the Kirchhoff-Helmholtz integral [27, 31]. In the present work, we are more interested in positioning the auditory event at any point in a virtual 3D space using a pair of loudspeakers arranged in a stereophonic set-up. This is achieved by audio systems based on Head-Related Transfer Functions (HRTF). HRTF-based systems are also able of creating multiple virtual sound images simultaneously at different positions in the same listening space using two transducers only. This chapter introduces the basic psychoacoustical, physical, and signal processing principles behind virtual sound source systems of this type. The chapter is divided into two parts: an analysis section (Section 2.1) and a synthesis section (Section 2.2). In Section 2.1, psychoacoustical and physical phenomena related to natural spatial hearing are introduced with emphasis on the factors that are exploited in improving the robustness of 3D sound systems in Chapter 4. The basic problem of virtual sound source synthesis using HRTFs is examined in Section 2.2.

2.1 Sound Localisation and Localisation Cues

In natural hearing, the human auditory system makes use of several cues to locate an auditory event in the listening space. The difference between sound signals at the left and right ears (interaural cues) are used to determine the horizontal component (azimuth angle) of the sound source position. Spectral cues resulting from high frequency reflections inside the listener's outer ears (pinnae) help determining the vertical component (elevation angle). Human listeners are also capable of estimating the sound source distance in addition to azimuth and elevation angles. Interaural, spectral, and distance cues are discussed in Sections 2.1.1, 2.1.2, and 2.1.3, respectively. When the above mentioned cues are ambiguous, humans usually move their heads (some animals move their ears), so that more (or less) interaural and spectral cues are introduced, making the localisation task easier. These dynamic cues are mentioned in Section 2.1.4. Room reverberation degrades human localisation performance, since the direction of a reflected sound may be confused with the direction of the sound coming directly from the source. However, thanks to the precedence effect, human listeners are still able to localise sounds in reverberant environments. The precedence effect and the physical and perceptual aspects of room reverberation are discussed in Sections 2.1.5 and 2.1.6, respectively. Most of the above mentioned cues are embedded in the physical transfer functions between a sound source and

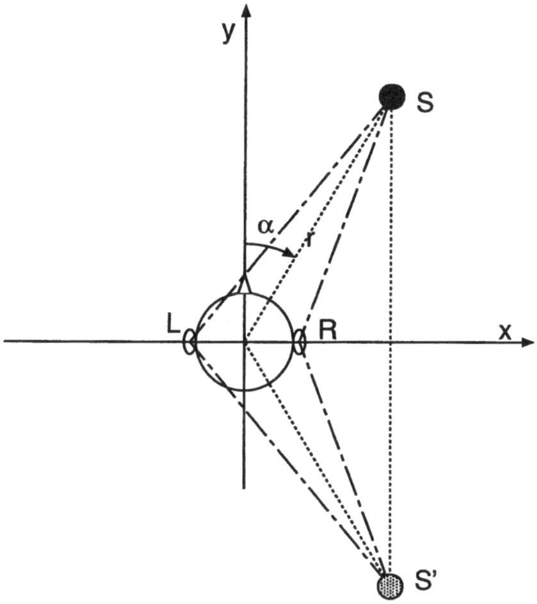

Figure 2.1: A sound source S and its image S' about the interaural axis introduce maximally similar interaural cues.

the eardrums of the listener. The properties of such transfer functions, referred to as the Head-Related Transfer Functions (HRTFs) are mentioned briefly in Section 2.1.7. Section 2.1.8 concludes the analysis part of this chapter with a brief review of the frequency selectivity of the human ears, a property that will be exploited in improving the robustness of 3D sound systems in Chapter 4.

2.1.1 Interaural Cues

In natural listening situations, the difference between sound waves received at the listener's left and right ears is an important cue used by the human auditory system to estimate the sound source position in space. These difference cues, referred to as the *interaural* or *binaural* cues are best explained using the far field anechoic listening situation shown in Fig. 2.1. A sound signal emitted from a sound source S located in the horizontal plane at azimuth angle α and distance r from the centre of the listener's head travels to the listener's right (ipsilateral) and left (contralateral) ears through paths SR and SL, respectively. Since SR in this example is shorter than SL, a tonal sound wave reaches the right ear before the left ear. The arrival time difference is referred to as the *Interaural Time Difference* (ITD). The ITD is zero when the source is at azimuth zero, and is a maximum at azimuth $\pm 90°$. This represents a difference of arrival time of about 0.7 ms for a typical-size human head, and is readily perceived [30]. ITD represents a powerful and dominating cue at frequencies below about 1.5 kHz. At higher frequencies, the ITD represents ambiguous cue since it corresponds to a shift of many cycles of the incident sound wave. For complex sound waves, the ITD of the envelope at high frequencies, which is referred to as the *Interaural Envelope Difference* (IED), is perceived.

On the other hand, the head forms an obstacle for incident sound waves. This leads to a level difference between the two ears, known as the *Interaural Intensity Difference* (IID). Besides being dependent on azimuth angle, the IID is highly dependent on the frequency of the incident sound. At low frequencies, the wavelength is larger than the listener's head and the sound wave is diffracted around the head to reach the contralateral ear without noticeable attenuation. As the frequency increases, the head forms a bigger obstacle for the sound wave and the level at the contralateral ear decreases. This is known as the head-shadow effect. The IID is an effective cue in the frequency range above 1.5 kHz, therefore, forming a complementary cue to the ITD. Together, the ITD and IID cover the whole audible frequency range.

Assuming a spherical head without any external ears, a sound source at azimuth angle α and its image about the interaural axis at azimuth $180° - \alpha$, as shown in Fig. 2.1, produce the same ITD and IID cues at the listener's ears. For a real head, the ITD and IID are never equal for the two positions, they are, however, maximally similar. With the absence of other spatial cues than the ITD and IID, this similarity causes potential confusion, which explains the often reported phenomenon of *front-back reversals* [25, 30]. This phenomenon is a special case, in the horizontal plane, of the general phenomenon of the *cones of confusion.* The cones of confusion are best explained using the vertical-polar coordinates mentioned in Section 1.5. In this coordinate system, every sound source at coordinates (r, α, θ) has an image source at $(r, 180° - \alpha, -\theta)$ that introduces maximally similar ITD and IID at the listener's ears.

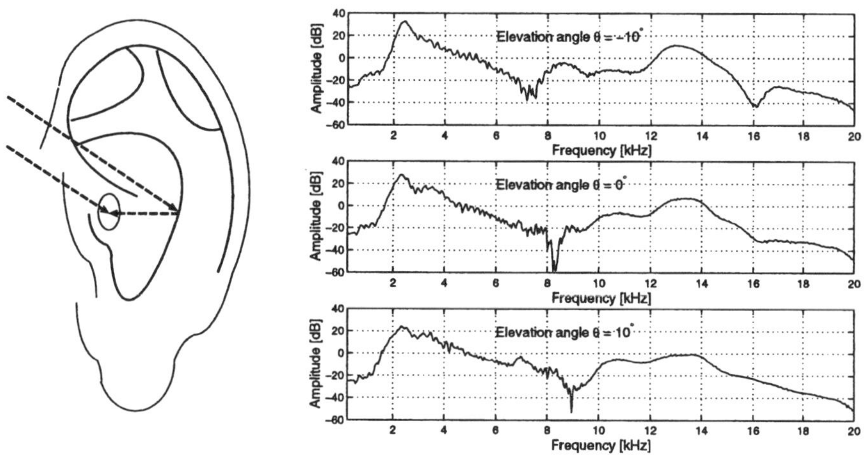

[a] [b]

Figure 2.2: [a] A schematic diagram of the high frequency reflections off the pinna causing constructive and destructive interferences with the direct sound wave. [b] measured pinna responses for a dummy head with the source in the median plane at elevation angles $-10°$, $0°$, and $10°$.

The spectral cues discussed in the next section enable resolving those confusions.

2.1.2 Spectral Cues

The primary cues used by the human auditory system for elevation estimation are often said to be *monaural*. This is in contrast with the *interaural* or *binaural* cues used for azimuth estimation. Spectral cues stem from reflections of short wavelength sound waves off the listener's upper body (torso) and off the outer ears (pinnae) [30]. The pinna with its irregular shape and resonant cavities acts as an acoustical filter. Sound waves reflected off the pinna interfere with the direct sound entering the ear canal constructively at some frequencies, and destructively at other frequencies as shown in Fig. 2.2-[a]. This leads to spectral peaks at frequencies where constructive interference occurs, and spectral dips at those frequencies where destructive interference takes place. The frequencies at which spectral peaks and dips appear, as well as the magnitude of these features, are dependent on the direction of the sound source. The frequencies at which spectral dips occur are of special interest, since spectral dips are often more localised in frequency than spectral peaks. The first spectral dip, known as *the pinna notch* is believed to be the major cue for elevation localisation. The frequency at which the pinna notch appears changes from about 6 to 12 kHz as the elevation angle changes from $-40°$ to $+60°$ [72] (see also Section 2.1.7). This is shown in Fig. 2.2-[b] for the transfer functions measured for a dummy head with the sound source in the median plane at elevation angles -10, 0, and 10 degrees[1]. From familiarity with their own pinnae response, human listeners are able to use the spectral cues to estimate the sound source position. Since spectral cues are mainly due to high frequency reflections, slight changes in the pinna shape may lead to significant changes in its frequency response. Therefore, spectral cues vary significantly among people due to differences in pinnae sizes and fine geometrical structure.

2.1.3 Distance Cues

Many phenomena have been noticed to influence the estimation of the distance of a sound source by the human auditory system. Loudness and direct-to-reverberant energy ratio are believed to be the most effective in influencing distance perception. Loudness cues stem from the fact that the sound pressure in the far field radiated from a sound source decreases with increasing the distance from the source. Therefore, nearby sound sources are perceived louder than distant sources emitting the same acoustic energy. The ratio of the sound intensity of two sources at distances r_1 and r_2 from a listener's ear is given by $I_1/I_2 = r_2^2/r_1^2$, which is known as the *inverse square law* [25]. Thus, a distance doubling decreases the sound intensity at the listener's ear by 6 dB. The human auditory system makes use of this fact in estimating the distance of a sound source. However, familiarity with the sound and its characteristics influence the estimated distance based on loudness [25]. Furthermore, the loudness cue is valid only in anechoic environments, since in a reverberant environment the sound distribution is dependent on the reverberation characteristics of the enclosed space. For instance,

[1] Data from `ftp://sound.media.mit.edu/pub/Data/KEMAR`. See also [73] for a description of the measurement procedure.

in a reverberant room, the sound field beyond the reverberation distance may be considered diffuse, and theoretically independent of the distance from the source (see Section 2.1.6). This explains the importance of direct-to-reverberant energy ratio as a distance cue [24, 30]. A recent study of distance perception [33] shows that a model based on a modified direct-to-reverberant energy ratio can accurately predict the source distance. The modification to the true direct-to-reverberant energy ratio is the use of an integration time window of 6 milliseconds in the calculation of the direct sound energy. Because of its importance in distance and environmental context perception as well as its influence on cross-talk cancellation (Section 2.2.2), reverberation and diffuse field characteristics are discussed in more details in Section 2.1.6.

2.1.4 Dynamic Cues

In ambiguous natural listening situations where interaural and spectral cues produce insufficient information for the auditory system to localise the sound source, humans turn their heads to resolve this ambiguity. This often occurs when ambiguous interaural cues are introduced at a listener's ears due to the cones of confusion phenomenon. A sound source at an azimuth angle α to the right of the listener in the horizontal plane introduces maximally similar interaural cues as a source at azimuth $180° - \alpha$ as mentioned in Section 2.1.1. A human listener would resolve the ambiguous interaural cues by turning his/her head to the right, since the ambiguous cues still suggest that the source is at the listener's right. After turning right, if the interaural cues are minimised, the listener would decide that the source is now in front, otherwise if it is maximised, the decision would be that the source is at the back. Head motions of this type are shown to improve human localisation and decrease front-back reversals [136, 137]. Although head movements improve localisation in natural hearing, they constitute great difficulties to synthetic 3D sound systems. The problem of head movements in synthetic 3D sound systems is discussed in details in Chapter 4.

2.1.5 The Precedence Effect

Natural sound localisation is affected by the above mentioned cues as well as by numerous other psychoacoustical phenomena. In this section, one of those phenomena, *the precedence effect*, that is directly related to localisation in reverberant environments, is briefly mentioned. The precedence effect, also known as *the law of the first wavefront* [30], explains an important inhibitory mechanism of the human auditory system that allows human listeners to localise sounds in reverberant environments. When the combination of direct and reflected sounds is heard, the listener does perceive the sound to be coming from the direction of the sound that arrives first at his ears. This is true even if the reflected sound is more intense than the direct sound [82]. However, the precedence effect does not totally eliminate the effect of a reflection. Reflections add a sense of spaciousness and loudness to the sound.

Experiments with two clicks of equal intensity have shown that if the second click arrives within 1 ms after the first, the two clicks are perceived as an integrated entity. The perceived location of this entity obeys the *summing localisation* regime [82]. Within this regime, there is an orderly weighting such that as the delay increases, the weighting decreases. For delays between 1 and 4 ms, the precedence effect is in operation and it is maximum at about 2 ms delay, where the sound location is perceived to be at the location of the first click. Finally, in the range between 5 to 10 ms, the sound is perceived as two separate clicks and the precedence effect starts to fail. However, it was noticed that the second click not only contributes to spaciousness but the perceived location is also biased towards the position of the second click. Furthermore, the second sound was found to decrease the accuracy of azimuth and elevation localisation compared to anechoic listening situations [24, 25].

In normal listening situations, sound signals last longer than clicks, and reflections arrive at the listener's ears while the direct signal is still heard. In such situations, the precedence effect operates on onsets and transients in the two signals. Furthermore, it was found that the precedence effect for speech signals, better known as the *Haas effect*, has much different time constants than those mentioned above [82]. Maximal suppression occurs for a delay between 10 to 20 ms, while speech comprehension is affected by reflections later than 50 ms.

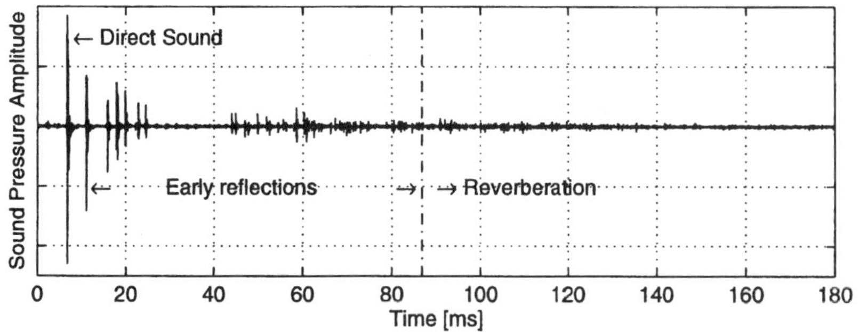

Figure 2.3: A calculated impulse response between two points in a room.

2.1.6 Reverberation and Room Acoustics

Spatial sound pressure distribution in an enclosure is rather irregular. Near sound absorbing surfaces, the sound pressure may decrease significantly. Near corners and rigid surfaces, constructive interference of reflections may cause local pressure peaks. Near a sound source, the direct sound dominates. Nevertheless, at areas far from the source, where the reflected sounds play a predominant role, the sound field is often considered as being diffuse, i.e. built up by equally strong uncorrelated plane waves from all directions. At a certain point in the field, the intensities of the waves incident from each direction are equal and the net energy transport in a pure diffuse field is zero [48]. A sound field can be approximated by the diffuse model only if reflections from several directions play a significant role, i.e. there are acoustically hard surfaces at all sides of the enclosure. In this section, a brief review of room acoustics and the diffuse field characteristics is given.

2.1.6.1 Room Impulse Response

The sound field distribution in a room or an enclosure can be calculated by solving the three-dimensional wave equation knowing the boundary conditions in terms of acoustic impedance or reflection coefficients. However, for a certain combination of source-sensor positions, the acoustic transmission can be described by the impulse response from a source to a sensor. Figure 2.3 shows such an impulse response calculated for a fictitious enclosure.[2] The impulse response shows the direct sound arriving first, followed by distinct reflections, known as the *early reflections* generated at the room boundaries. Early reflections may be deterministically calculated by solving the eigenfunction problem inside the room for low order eigenmodes. These reflections increase in density and decrease in amplitude with time. After about 80 ms from the direct sound, the reflections arrive equally from all directions and the process is best described statistically as exponentially decaying noise. The statistical part of the impulse response is often referred to as the *late reverberation.* An important property of the impulse response between two points in a room is that if the source or sensor position changes slightly, the fine structure of the impulse response may change significantly. This is explained by the wide range of wavelengths covered by the audio band and the low speed at which the sound propagates in the air. The audio band covers wavelengths from about 17 mm at 20 kHz to 17 m at 20 Hz. Therefore, a small displacement of a few millimetres corresponds to a large part of a cycle at high frequencies while at low frequencies such a displacement corresponds to a negligible fraction of a cycle. This spatial-frequency dependency of a room impulse response makes it difficult to control the sound field in large zones in a wide frequency range as will be shown in Chapter 4.

2.1.6.2 Reverberation Time

A very important parameter in room acoustics is the reverberation time. The reverberation time T_{60} is defined as the time needed for the sound pressure level to decay by 60 dB when a steady state sound source in the room is suddenly switched off. An approximate formula for the reverberation time is $T_{60} \approx V/6\bar{\beta}S$, where V is the room volume in m^3, $\bar{\beta}$ is the average absorption coefficient of the room boundaries, and S is the surface area of the room in m^2. Since the average absorption coefficient $\bar{\beta}$ is frequency dependent, the reverberation time is also frequency dependent, and is usually given as the average value in an octave band.

At low frequencies, an enclosure may be considered as a three dimensional resonator. The sound waves reflected from the enclosure bound-

[2]The image source program "Room Impulse Response 2.1" [71] has been used to calculate this impulse response.

aries constitute standing wave patterns or *eigenmodes* and resonance may occur for a large number of *eigenfrequencies.* The number of these eigenmodes increase rapidly with frequency and beyond a certain frequency, *the Schroeder frequency,* very large number of modes are excited and the sound field may be considered diffuse. The Schroeder frequency depends only on the enclosure volume V and the reverberation time T_{60} and can be approximated by the empirical formula $f_{sh} \approx 2000\sqrt{T_{60}/V}$, where V is in m^3, and T_{60} is in seconds. The diffuse approximation is, therefore, valid at lower frequencies in larger or more reverberant rooms. For example, in a room of dimensions $V = 4 \times 6 \times 3\ m^3$, and a reverberation time $T_{60} = 0.5$ s, the sound field may be approximated by the diffuse field model for frequencies above 167 Hz. Below this frequency, the eigenmode model may be used to calculate the individual modes and their eigenfrequencies.

2.1.6.3 Reverberation Distance

The reverberation distance is an indication for the distance from the sound source beyond which the sound field may be considered diffuse. The direct sound pressure level L_d is dependent only on the source characteristics and the source-receiver distance, therefore, it decreases by 6 dB per distance doubling. When the direct sound meets the enclosure boundaries, a fraction of the acoustic energy is reflected to build the reverberation field. When the reverberation field is a pure diffuse field, the reverberant sound pressure level L_r is independent of the distance from the sound source. The reverberation distance r_r is then defined as the distance from the sound source at which the direct sound pressure and the reverberant sound pressure are equal. The reverberation distance may be approximated by $r_r = 0.25\sqrt{\beta S/\pi} \approx 0.06\sqrt{V/T_{60}}$. Fig. 2.4 shows the direct (L_d), reverberant (L_r), and total (L_t) sound pressure levels as functions of the distance r from the sound source. At distances close to the sound source ($r/r_r < 1$), the direct sound L_d dominates. At the reverberation distance, the total level L_t is 3 dB above the reverberation level. At distances beyond $3r_r$, L_r exceeds L_d by more than 10 dB and the direct sound may be neglected compared to the reverberant sound. In real rooms, however, the sound level tends to decrease slightly with increasing distance beyond the reverberation distance, and the sound field may be considered diffuse only by approximation.

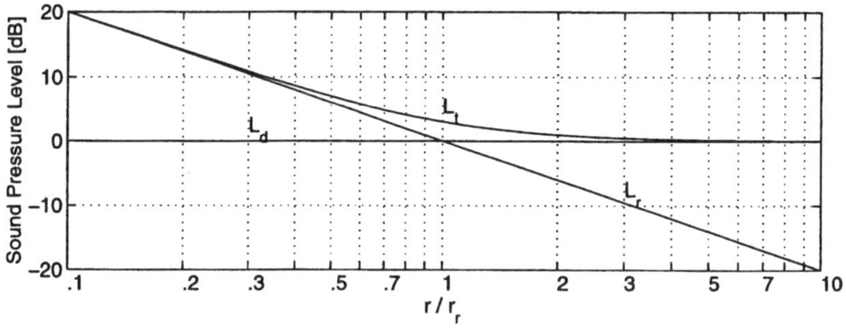

Figure 2.4: Direct (L_d), reverberant (L_r), and total (L_t) sound pressure levels in an enclosure as functions of the distance from the sound source.

2.1.6.4 Sound Localisation and Reverberation

In a reverberant environment and beyond the reverberation distance, the total sound pressure level is dominated by the reverberant field rather than by the direct field as shown in Fig. 2.4. Since the loudness cue for distance estimation is based on the decrease in sound pressure with increasing distance from the source as mentioned in Section 2.1.3, it is clear that the loudness cue is valid only in anechoic environments. Beyond the reverberation distance, which may be less than one meter in average rooms, the total sound pressure level is almost constant, and the loudness cue disappears. At distances smaller than the reverberation distance, the loudness cue may be considered an effective localisation cue.

As the loudness cue becomes less effective with increasing reverberation, the direct-to-reverberant energy ratio D/R becomes a more effective cue in distance perception as mentioned in Section 2.1.3. This ratio can be shown to be

$$\frac{D}{R} = \frac{(P_D)_{rms}}{(P_R)_{rms}} = \frac{r_r}{r}, \tag{2.1}$$

which is dependent only on the reverberation distance r_r, a characteristic of the diffuse field in the enclosure, and the distance r from the sound source. Therefore, D/R is considered to be a much more effective distance cue than the loudness cue in reverberant environments. Furthermore, reverberation is considered to be important for the perception of environmental context. The reverberation time and level together with the experience with sounds in reverberant rooms enable a listener to estimate the size and absorptiveness of the surfaces in the environment.

Although reverberation is important for distance and environmental context perception, it was found that localisation accuracy of azimuth and elevation degrades due to reverberation [24, 25]. This is explained by the ability of humans to detect the direction of the early reflections in severe reverberation conditions. The precedence effect mentioned in Section 2.1.5 only partially suppresses the effects of reflected sounds. Moreover, the reverberation makes it difficult for the auditory system to correctly estimate the ITD at low frequencies. This is because in typical rooms, the first reflections arrive before one period of a low frequency sound is completed. Thus, in a reverberant room, low frequency information is essentially useless for localisation and azimuth localisation is severely degraded. In such cases, the important timing information comes from the IED, e.g. from the transients at the onset of a new sound.

2.1.7 Head-Related Transfer Functions

Consider the natural listening situation where a sound signal is emitted from a source located at (r, α, θ) from the centre of the listener's head. The sound pressure at the listener's ears can be modelled by the convolution between the sound signal and the acoustic transmissions between the sound source and the listener's eardrums. The acoustic transmissions are defined by the impulse responses measured from the source to the listener's left and right ears. The sampled versions of such acoustic transmissions may be expressed as $\underline{\mathbf{h}}_L(n, r, \alpha, \theta)$ and $\underline{\mathbf{h}}_R(n, r, \alpha, \theta)$, respectively. This pair of impulse responses is often referred to as the *Head-Related Impulse Response* (HRIR) pair. The Discrete Fourier Transform (DFT) of $\underline{\mathbf{h}}_L$, $\underline{\mathbf{H}}_L(\omega, r, \alpha, \theta)$, and of $\underline{\mathbf{h}}_R$, $\underline{\mathbf{H}}_R(\omega, r, \alpha, \theta)$, are known as the *Head-Related Transfer Function* (HRTF) pair. An HRTF measured from the source to the listener's eardrum captures all the physical cues to source localisation. Although HRTFs are functions in four variables, for distances greater than about one meter in an anechoic environment, the source is in the far field, and the response falls inversely with the range as mentioned in Section 2.1.6. Most HRTF measurements are anechoic far field measurements, which reduces an HRTF to be a function of three variables, namely, azimuth, elevation, and frequency.

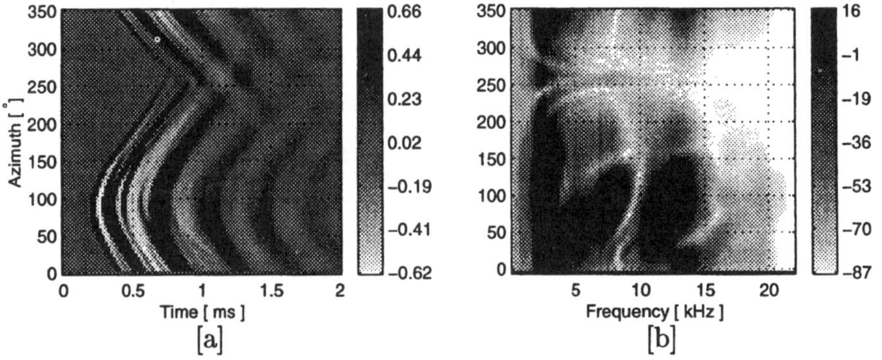

Figure 2.5: Measured HRTFs of KEMAR's right ear for a source in the horizontal plane. [a] is the amplitude of the HRIR and [b] is the amplitude of the HRTF in dB.

Usually HRTFs are measured in an anechoic environment, and thus do not include the effects of reverberation, which is important for range estimation and environmental context perception. In that case, unless binaural room simulation is used to introduce these important reflections, an improper ratio of direct-to-reverberant sound energy results, and when heard through headphones, the sound often seems to be either very close or inside of the head. It is possible to measure the HRTFs in an actual reverberant setting. This has the disadvantages of limiting the simulated virtual environment to a particular room and leads to very long impulse responses. In the following chapters, both anechoic and reverberant HRTFs will be used.

Anechoic HRTFs of manikins and human subjects have been intensively studied in search for physical characteristics that are related to sound localisation. In the subsequent chapters, frequent use will be made of an anechoic HRTF set measured for the KEMAR manikin [72, 73]. The impulse response of the right ear of KEMAR in the horizontal plane as a function of azimuth angle is shown in Fig. 2.5-[a]. The interaural cues can be readily recognised in Fig. 2.5-[a]. The sound has the highest amplitude and arrives first when it is coming from the right side (azimuth = 90°). Conversely, it has the lowest amplitude and arrives latest when it is coming from the left side (azimuth = 270°). The arrival time varies with azimuth in a more or less sinusoidal fashion as estimated by a spherical head model [30, 54]. Pinna reflections can also be seen in the initial

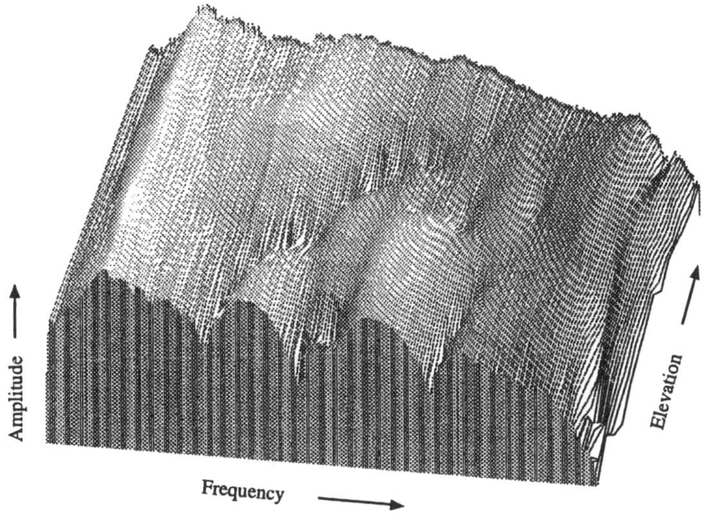

Figure 2.6: The amplitude in dB of the HRTF of KEMAR's right ear against linear frequency for a source in the median plane spanning elevation angles from $-40°$ to $90°$.

sequence of rapid changes when the source is located at the right side of the head. The cones of confusion phenomenon can also be expected as the response is almost symmetrical about horizontal lines at azimuth = 90° and azimuth = 270°, which constitute the interaural axis.

Fig. 2.5-[b] shows the Fourier transform of the impulse responses in Fig. 2.5-[a]. It is also clear from Fig. 2.5-[b] that the response is highest when the source is at the right and weakest when the source is at the left. The pinna notch is also visible around 10 kHz when the source at the right side of the head. For the opposite side, the sound pressure is low due to head shadowing head-shadow effect, and the notch is not clear. The broad peak in the range 2 to 3 kHz that is independent of the azimuth can be attributed to the ear canal resonance [72].

The frequency response of the right ear of KEMAR when the source spans the elevation angles between -40 and 90 degrees in the median plane as a function of elevation angle is shown in Fig. 2.6. Fig. 2.2-[b] shows three distinct cross-sections in Fig. 2.6 at elevations -10, 0, and 10 degrees, from which the exact frequency and amplitude scales can be seen. Unlike the response in the horizontal plane, interaural cues in the median plane are negligible. The effect of the pinna reflections is clearly visible in the spectral peaks and notches. The frequency of the first notch and its magnitude are elevation dependent, therefore, this notch is considered a major factor in elevation estimation. The frequency of the first notch ranges from 6 to 12 kHz as the elevation angle changes from $-40°$ to $60°$. For elevation angles above about $60°$, the notch disappears and there is no spectral dependency on elevation. The ear canal resonance is also visible as the first broad spectral peak which is independent of the elevation angle.

2.1.8 Frequency Selectivity of Human Ears

The fundamental characteristics of the human auditory system can be described in a high-level functional model as a bank of overlapping linear band-pass filters. Psychoacoustical experiments have been used to study the characteristics of this auditory spectral analyser. These experiments led to two main models: the *Critical Bands* (CB) model [148] and the *Equivalent Rectangular Bandwidth* (ERB) model [116]. These two models are briefly described below.

2.1.8.1 Critical Bands Model

When the bandwidth of a band-pass noise centred at a frequency f_c is increased while keeping the total energy constant, the loudness remains constant as long as the bandwidth of the band-pass noise is smaller than a certain value. Increasing the noise bandwidth beyond this bandwidth leads to increasing loudness. The critical bandwidth at the centre frequency f_c is defined as that bandwidth at which the loudness starts to increase [148]. The critical bands as a function of frequency may be determined by using the above mentioned loudness experiment. This loudness experiment is just one of five methods described by Zwicker [148] to measure the critical bandwidth. Collecting data from several experiments, it was found that the critical bands show constant bandwidth of 100 Hz for frequencies below 500 Hz, while at frequencies above, the critical bands are about 20% of the centre frequency. This is approximated by

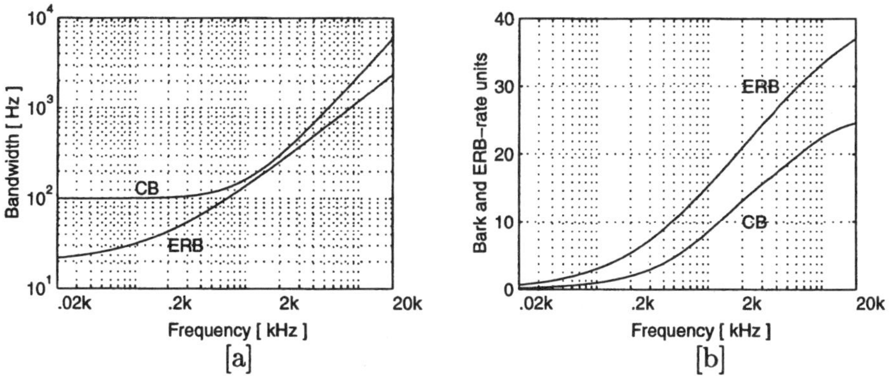

Figure 2.7: [a] Critical Bands and Equivalent Rectangular Bandwidth as functions of frequency. [b] Bark and ERB-rate scales as functions of frequency.

$$\Delta f_{CB} = 25 + 75\,[1 + 1.4\,f^2]^{0.69}, \tag{2.2}$$

where f is in kHz and Δf_{CB} in Hz. This relationship is plotted in Fig. 2.7-[a].

The critical bands concept is so important that it has been defined as a perceptual scale, the Bark scale, to replace the physical frequency scale. This is done by considering each critical band as a unit on the new scale and, therefore, the audio frequency band up to 16 kHz is divided into 24 critical bands or Barks. One Bark on the Bark scale corresponds to a constant distance of about $1\frac{1}{3}$ mm along the basilar membrane and, therefore, matches the frequency-place transformation that takes place in the inner ear (cochlea) [148]. The relationship between the Bark scale and the frequency scale is shown in Fig. 2.7-[b] and is expressed mathematically by [148]

$$z = 13\,\arctan(0.76\,f) + 3.5\,\arctan(f/7.5)^2, \tag{2.3}$$

where z is the Bark value and f is the frequency in kHz.

2.1.8.2 Equivalent Rectangular Bandwidth Model

In the Equivalent Rectangular Bandwidth (ERB) model, the auditory system is considered as a bank of band-pass filters. The shape of the auditory filter at a given centre frequency is measured using a notched-noise masking experiment [116]. This experiment is performed by measuring the threshold of a tone at the centre frequency in question, masked by white noise with a notch (band-stop) centred at the tone frequency, as a function of the notch bandwidth. The auditory filter shape is then the derivative of the threshold curve with respect to the normalised frequency variable $g = \mid f - f_c \mid / f_c$. From these experiments, the auditory filter shape could be approximated by a rounded-exponential filter that takes one of three forms depending on the range over which the filter needs to be modelled. The simplest form of such filter is given by [116]

$$W(g) = (1 + pg)\, e^{-pg}, \tag{2.4}$$

where $W(g)$ is the auditory filter transfer function, g is the normalised frequency given above, and p is a parameter that controls the bandwidth and the attenuation rate of the filter. The bandwidth of the auditory filter (2.4) centred at f_c is equivalent to the bandwidth of a rectangular filter centred at the same frequency and passing the same noise power. Therefore, the bandwidth of an auditory filter is called the Equivalent Rectangular Bandwidth (ERB) at that specific frequency. The ERB corresponds to a constant distance of about 0.85 mm on the basilar membrane, and is related to frequency by

$$ERB = 19.5\,(6.046\, f + 1), \tag{2.5}$$

where ERB is in Hz and f is in kHz. This equation is plotted on the same graph with the critical bandwidth in Fig. 2.7-[a]. The main difference between the two curves is seen at low frequencies. While the CB model shows constant bandwidth at frequencies below 500 Hz, the ERB model indicates that the bandwidth is frequency dependent in this frequency range. Although many recent experiments suggest that the ERB model is more accurate than the classical CB model, both models are widely used in practical applications. The CB model is especially more used in audio coding applications [80].

Similar to the critical bands, the ERB has also been taken as a perceptual scale replacing the frequency scale, and leading to the ERB-rate scale. The ERB-rate scale relates the number of ERBs to frequency in kHz in the same way as the Bark scale relates the number of CBs to frequency. This relationship is shown in Fig. 2.7-[b] and is given mathematically by

$$\mathrm{ERB-rate} = 11.17 \ln \left| \frac{\mathrm{f} + 0.312}{\mathrm{f} + 14.675} \right| + 43. \tag{2.6}$$

Regardless of the discrepancy between the two models at low frequencies, both models suggest that the spectral filtering in the human auditory system can be modelled as a non-uniform filterbank. The bandwidths of the filters increase with increasing centre frequency in a constant-Q (percentage bandwidth) manner. This suggests that human ears perform multiresolution spectral analysis. Lower frequencies are analysed with higher frequency resolution (and, therefore, lower time resolution) than high frequencies. It may also be interpreted as spectral averaging with wider averaging windows at higher frequencies. This property of increasingly coarser resolution with increasing frequency will be exploited in Chapter 4 in improving the robustness of 3D sound systems.

2.2 Synthesis of Virtual Sound Sources

The problem of synthesis of virtual sound sources is concerned with creating an auditory event at an arbitrary point in a virtual 3D space. A perfect virtual source synthesis system must faithfully emulate all the localisation cues mentioned in Section 2.1, so that the listener perceives the sound image as a convincing real sound source. This ideal synthesis is difficult to achieve and in most cases simplifications and approximations are made to make the system realisable. For instance, the dynamic cues are often ignored and the distance cues are implemented using binaural room simulation or artificial reverberation. Furthermore, HRTFs for dummy heads are often used to create the interaural and spectral cues instead of the listener's own HRTFs. This section reviews the basic principles of HRTF-based virtual sound source synthesis. Section 2.2.1 introduces the basic principles of HRTF-based 3D sound systems that use headphones to deliver their output sounds to their users. Systems that deliver their output sounds through loudspeakers are discussed in Section 2.2.2.

2.2.1 Binaural Synthesis and Headphone Displays

As mentioned in Section 2.1.7, a natural listening situation may be modelled by filtering a monophonic sound signal $x(n)$ by a pair of Head-Related Impulse Responses (HRIR) $\underline{\mathbf{h}}_L(n,r,\alpha,\theta)$ and $\underline{\mathbf{h}}_R(n,r,\alpha,\theta)$ corresponding to a sound source at a position (r,α,θ). Conversely, filtering a monophonic sound signal through the HRIR pair measured for a sound source at an arbitrary point (r, α, θ) in a 3D space creates an auditory event at that point. This is expressed in the frequency domain at a frequency ω by (see Section 1.5 for notation conventions)

$$\underline{\mathbf{E}}(\omega, r, \alpha, \theta) = \underline{\mathbf{H}}(\omega, r, \alpha, \theta)\, X(\omega), \tag{2.7}$$

where $\underline{\mathbf{E}}(\omega, r, \alpha, \theta) = [E_L(\omega, r, \alpha, \theta) \;\; E_R(\omega, r, \alpha, \theta)]^T$ is a column vector of left and right ear signals expressed at ω, and usually referred to as the *binaural signals*. $\underline{\mathbf{H}}(\omega, r, \alpha, \theta) = [H_L(\omega, r, \alpha, \theta) \;\; H_R(\omega, r, \alpha, \theta)]^T$ is a column vector of the HRTF pair, and $X(\omega)$ is the monophonic sound signal at ω that is independent of the spatial coordinates. Provided that the HRIR pair matches those of the listener, playing the binaural signals at the corresponding listener's eardrums (e.g. through headphones) creates an auditory event (a virtual source) at (r, α, θ). In general, K auditory events may be created simultaneously in the same virtual space by extending the scalar sound signal in (2.7) to be a column vector of length K. Omitting the dependency on spatial coordinates, this can be expressed as

$$\begin{bmatrix} E_L(\omega) \\ E_R(\omega) \end{bmatrix} = \begin{bmatrix} H_{L_1}(\omega) & \cdots & H_{L_k}(\omega) & \cdots & H_{L_K}(\omega) \\ H_{R_1}(\omega) & \cdots & H_{R_k}(\omega) & \cdots & H_{R_K}(\omega) \end{bmatrix} \begin{bmatrix} X_1(\omega) \\ \vdots \\ X_k(\omega) \\ \vdots \\ X_K(\omega) \end{bmatrix}, \tag{2.8}$$

which may be expressed compactly as

$$\underline{\mathbf{E}}(\omega) = \mathbf{H}(\omega)\, \underline{\mathbf{X}}(\omega). \tag{2.9}$$

While (2.8) gives the binaural signals at one frequency only, it should be kept in mind that there are as many equations of the form (2.8) as there are frequencies. Using frequency representations of signals at a single frequency effectively reduces the matrix $\mathbf{H}$ from an array of 3-indexes (ear, input signal, frequency) to a matrix of 2-indexes (ear, input signal). This enables using well known matrix algebra while keeping good understanding of the physical process. In the following, the explicit dependency on ω will also be dropped to enhance the readability of the equations.

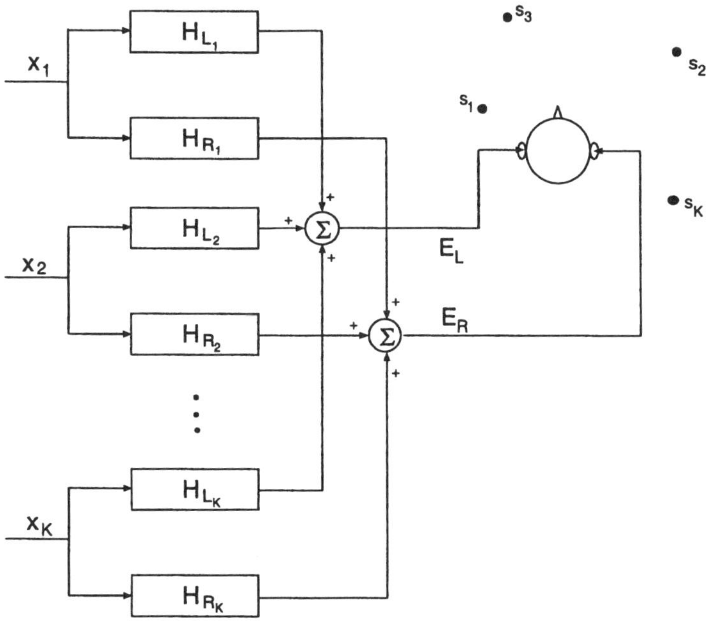

Figure 2.8: Synthesis of multiple virtual sound sources at the ears of a single listener using headphones.

The element X_k of the vector $\underline{\mathbf{X}}$ in (2.9) corresponds to the sound signal played by the kth virtual source and, therefore, must be filtered by the HRTF pair $\underline{\mathbf{H}}_k = [H_{L_k}\ H_{R_k}]^T$. This is shown in a block diagram form in Fig. 2.8. In this case, only two transducers are required to deliver the binaural signals to the corresponding listener's ears. This justifies the statement made in the introduction of this chapter that K virtual sound sources may be created using two transducers.

The above principle can be further generalised to create K virtual sound images at the ears of M listeners by expanding the column vector $\underline{\mathbf{E}}$ and the HRTF matrix $\mathbf{H}$ to be of dimensions $[2M \times 1]$ and $[2M \times K]$, respectively, which results in

$$\begin{bmatrix} E_{L_1} \\ E_{R_1} \\ E_{L_2} \\ E_{R_2} \\ \vdots \\ E_{L_M} \\ E_{R_M} \end{bmatrix} = \begin{bmatrix} H_{L_{11}} & H_{L_{12}} & \cdots & H_{L_{1K}} \\ H_{R_{11}} & H_{R_{12}} & \cdots & H_{R_{1K}} \\ H_{L_{21}} & H_{L_{22}} & \cdots & H_{L_{2K}} \\ H_{R_{21}} & H_{R_{22}} & \cdots & H_{R_{2K}} \\ \vdots & \vdots & \ddots & \vdots \\ H_{L_{M1}} & H_{L_{M2}} & \cdots & H_{L_{MK}} \\ H_{R_{M1}} & H_{R_{M2}} & \cdots & H_{R_{MK}} \end{bmatrix} \begin{bmatrix} X_1 \\ X_2 \\ \vdots \\ X_K \end{bmatrix}. \tag{2.10}$$

In (2.10), E_{L_m} and E_{R_m} are the signals at the left and right ears of the mth listener, respectively, and $H_{L_{mk}}$ and $H_{R_{mk}}$ form the HRTF pair measured from the ears of the mth listener to the position of the kth source. In this case, M headphones are required, one for each listener.

In 3D sound systems, the question arises of how to deliver the electrical binaural signals to the listener's eardrums as acoustic waves. Headphones deliver E_L at the left ear without any cross-talk from E_R and E_R at the right ear without any cross-talk from E_L and, therefore, form a simple way of performing this transducer function. However, headphones have their own drawbacks: they may not be comfortable to use for a long period of time. They also isolate their user from the surrounding environment. Sounds heard over headphones often seem to be too close or inside the listener's head. Since the physical sources (the headphones) are actually very close to the listener's ears, compensation is needed to eliminate the acoustic cues to their locations. This compensation is sensitive to the headphones position. Finally, headphones can have notches and peaks in their frequency responses that resemble the pinna responses. If uncompensated headphones are used, elevation perception may be severely compromised [54].

2.2.2 Loudspeaker Displays and Cross-Talk Cancellation

Alternatively, a pair of loudspeakers arranged in a stereophonic setting may be used to transform the binaural signals to acoustic waves. Such 3D sound systems are often referred to as *loudspeaker displays*. In loudspeaker displays, the sound emitted from each loudspeaker is heard by both ears. Furthermore, on its way to the listener's ears, the sound is filtered through a mixing matrix of the four acoustic transfer functions

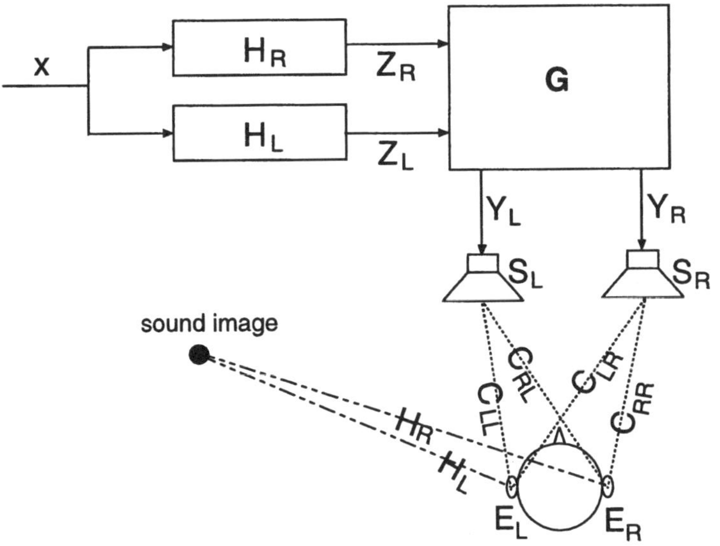

Figure 2.9: Synthesis of a single virtual sound source using two loudspeakers.

between the loudspeakers and the ears as shown in Fig. 2.9. To correctly deliver the binaural signals to the listener's eardrums, this mixing process has to be inverted. This inversion is usually performed by a network of filters known as the *cross-talk canceller*, which is represented by the block **G** in Fig. 2.9. Using frequency domain representations of signals and dropping the frequency and spatial dependency for clarity, the sound waves at the left and right ears can be expressed as

$$\begin{bmatrix} E_L \\ E_R \end{bmatrix} = \begin{bmatrix} C_{LL} & C_{LR} \\ C_{RL} & C_{RR} \end{bmatrix} \begin{bmatrix} G_{11} & G_{12} \\ G_{21} & G_{22} \end{bmatrix} \begin{bmatrix} H_L \\ H_R \end{bmatrix} X, \tag{2.11}$$

which may be written compactly as $\underline{\mathbf{E}} = \mathbf{C}\,\mathbf{G}\,\underline{\mathbf{H}}\,X$. To faithfully deliver the binaural signals Z_L and Z_R to the corresponding ears, the cross-talk canceller **G** must equal the inverse of the matrix **C**, so that $\mathbf{CG} = \mathbf{I}$, and **G** is given by

$$\mathbf{G} = \frac{1}{C_{LL}C_{RR} - C_{RL}C_{LR}} \begin{bmatrix} C_{RR} & -C_{LR} \\ -C_{RL} & C_{LL} \end{bmatrix}. \tag{2.12}$$

The elements of the square matrix **C** are the acoustic transfer functions between the listener's eardrums and the loudspeakers as shown

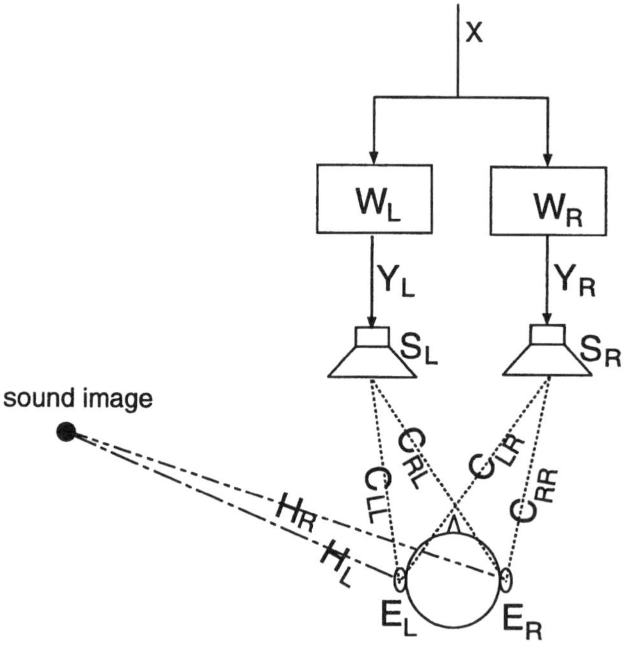

Figure 2.10: A loudspeaker display system employing a combined binaural synthesis and cross-talk cancellation filters.

in Fig. 2.9. Each of these transfer functions C_{AB} is composed of the responses of the corresponding loudspeaker S_B, the head-related transfer function H_{AB}, the room between the corresponding loudspeaker and ear R_{AB}, the microphone used to measure the sound at the eardrum M_A, the Analogue-to-Digital Converter (ADC), the Digital-to-Analogue Converter (DAC), and the signal conditioning devices. Therefore, the elements of **C** are complex functions in frequency and space coordinates. They are often non-minimum phase and contain deep notches due to pinnae reflections and room reverberation. Moreover, since loudspeakers are acoustic band-pass filters, their responses have much less energy at the edges of the audio frequency band. For those reasons, the inversion of **C** required by the cross-talk cancellation matrix in (2.12) is difficult to calculate. This is in addition to the processing complexity required to implement the matrix inversion in real-time.

A computationally attractive approach to loudspeaker displays is to combine the binaural synthesis and the cross-talk cancellation functions as shown in Fig. 2.10. In this configuration, the [2-input × 2-output] cross-talk canceller, which requires four separate filters as given by (2.12) is combined with the [1-input × 2-output] binaural synthesis subsystem. The combined solution is, therefor, [1-input × 2-output] and can be implemented with two filters given by $\underline{\mathbf{W}} = \mathbf{C}^{-1}\,\underline{\mathbf{H}}$, and expressed fully by

$$\begin{bmatrix} W_L \\ W_R \end{bmatrix} = \frac{1}{C_{LL}C_{RR} - C_{RL}C_{LR}} \begin{bmatrix} C_{RR}H_L - C_{LR}H_R \\ -C_{RL}H_L + C_{LL}H_R \end{bmatrix}. \tag{2.13}$$

This approach does not *explicitly* require a matrix inversion and embeds the convolution with the HRTF pair in the filters' coefficients. Therefore, it is computationally more efficient than the separate implementation of binaural synthesis followed by cross-talk cancellation. Furthermore, the convolution with the HRTFs, H_R and H_L, may compensate for some of the deep notches in the elements of $\mathbf{C}$, which may improve the numerical stability of the total solution, although $\mathbf{C}^{-1}$ may be ill-conditioned. Moreover, it can be shown that a causal implementation of $\underline{\mathbf{W}}$ exists provided that $\underline{\mathbf{H}}$ possesses sufficient delay. However, the problem of inverting a non-minimum phase system remains and will be discussed in Chapter 3.

Similar to the binaural synthesis case, a loudspeaker display system may be generalised to accommodate more virtual sound images at the ears of multiple listeners. By analogy with (2.10), to create K virtual sound images at the ears of M listeners, $\underline{\mathbf{E}}$ and $\mathbf{H}$ are extended to be of dimensions $[2M \times 1]$ and $[2M \times K]$, respectively. Unlike in binaural synthesis, the number of loudspeakers is not constrained to $2M$ since all listeners share listening to the loudspeakers as long as they share the listening space. Let the number of reproduction loudspeakers be L, then the dimension of $\mathbf{C}$ becomes $[2M \times L]$. For the signals at the listeners' ears to become equal to the natural listening situation with L loudspeakers and K sources, the following equation must be valid at every frequency ω

$$\underline{\mathbf{E}}(\omega) = \mathbf{H}(\omega)\,\underline{\mathbf{X}}(\omega) = \mathbf{C}(\omega)\,\mathbf{W}(\omega)\,\underline{\mathbf{X}}(\omega), \tag{2.14}$$

where $\mathbf{W}$ is an $[L \times K]$ matrix of control filters. The solution to those filters is then given by

$$\mathbf{W}(\omega) = \mathbf{C}^{-1}(\omega)\ \mathbf{H}(\omega). \tag{2.15}$$

The existence of the solution (2.15), the choice of the number of reproduction loudspeakers L, and the implementation details of multichannel loudspeaker displays are discussed at length in Chapter 3.

Chapter 3

Adaptive Filters Approach to 3D Sound Systems

The control filters in loudspeaker displays must implement both binaural synthesis and cross-talk cancellation functions as discussed in Section 2.2. The cross-talk cancellation subsystem is an inverse of an often non-minimum phase and ill-conditioned matrix of electro-acoustic transfer functions. An exact solution to the cross-talk canceller is difficult to calculate and in some cases, may not be possible at certain frequencies. Alternative to exact direct inversion, a statistical least mean square solution may be obtained for the matrix inverse. The advantage of such an approach is that it can be made adaptive, so that the filters can be designed *in-situ*. Such an adaptive solution also offers the possibility to track and correct the filters when changes in the electro-acoustic system occur. This chapter discusses theoretical and implementation issues concerning adaptive 3D sound systems.

An adaptive loudspeaker display that creates a single sound image at the ears of one listener requires two loudspeakers and two microphones placed inside the listener's ears. More microphones are needed as the number of listeners increases, and more filters are needed as the number of virtual sound images increases. Instead of treating any of the above special cases, a generalised model is introduced in Section 3.1. This model describes the general task of controlling the sound field produced by playing K pre-recorded audio signals through L loudspeakers at the proximity of M microphones. In addition to generalising the number of sound sources, loudspeakers, and microphones used in the reproduction system, the model is capable of describing a wide class of applications including synthesis of virtual sound sources, cross-talk cancellation, and active noise control.

The optimum least mean square solution for the control filters in the generalised model is derived in Section 3.2. Although important from the numerical analysis point of view, real-time implementation of the system would require approaching the optimum solution using an iterative algorithm. Such an iterative algorithm, the Multiple Error Filtered-X Least Mean Square (MEFX) algorithm, is introduced in Section 3.3.1. In Section 3.3.2, the convergence properties of the MEFX algorithm are examined, which shows the dependency of the convergence speed on the electro-acoustic transfer functions between the loudspeakers and the microphones. The effects of those electro-acoustic transfer functions are studied in detail in Section 3.3.3. This leads to an understanding of the consequences of some geometrical arrangements of the loudspeakers relative to the microphones.

Sound waves travelling from the reproduction loudspeakers arrive at the microphones after a considerable time delay due to the slow sound velocity in air. A direct consequence of this delay is that, in many cases, the solutions for the system filters become non-causal. This problem and its remedy are discussed in Section 3.3.4. Several other implementation details are discussed by considering the implementation of the MEFX algorithm in active noise control, virtual sound source synthesis, and cross-talk cancellation applications in Sections 3.3.5, 3.3.6, and 3.3.7, respectively.

In reverberant acoustic environments, the impulse response between two points may last for several hundreds of milliseconds. At the standard audio compact disc sampling frequency of 44.1 kHz, thousands of FIR filter coefficients are needed to properly model and store such an impulse response. Processing many of these transfer functions requires a huge amount of computations. Reducing the system complexity is, therefore, essential for real-time implementation. Efficient implementations using the Adjoint LMS and its Block Frequency Domain version are discussed in Section 3.4.

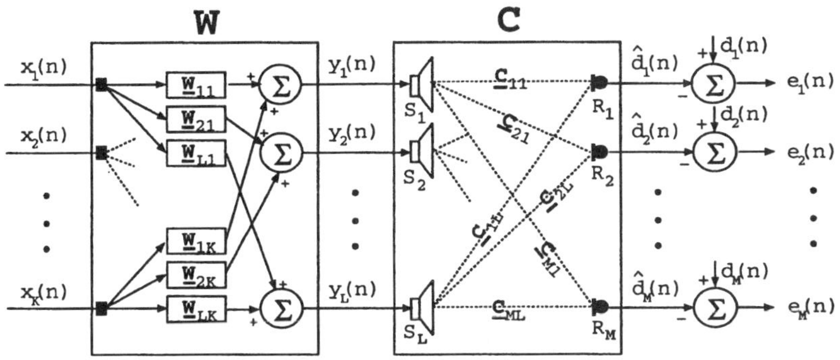

Figure 3.1: A generalised block diagram of a multichannel audio reproduction system.

3.1 A Generalised Model

3.1.1 Block Diagram of the Generalised Model

A generalised block diagram of a multichannel audio reproduction system is shown in Fig. 3.1. A set of L reproduction loudspeakers $\{S_1, S_2, \cdots, S_L\}$ are used to play K pre-recorded audio signals defined at the sample index n by

$$\underline{\mathbf{x}}(n) = \left[\ x_1(n) \quad x_2(n) \quad \cdots \quad x_K(n)\ \right]^T. \tag{3.1}$$

The sound field generated due to these loudspeakers is required to be controlled at the proximity of a set of M receivers $\{R_1, R_2, \cdots, R_M\}$ to a set of desired values defined as

$$\underline{\mathbf{d}}(n) = \left[\ d_1(n) \quad d_2(n) \quad \cdots \quad d_M(n)\ \right]^T. \tag{3.2}$$

Sound waves emitted from the loudspeakers are filtered through the $[M \times L]$ matrix of electro-acoustic transfer functions $\mathbf{C}(\omega)$ before reaching the microphones' positions. The matrix $\mathbf{C}(\omega)$ contains the Fourier transforms of the impulse responses $\{\underline{\mathbf{c}}_{ml} : m = 1, 2, \cdots, M,\ l = 1, 2, \cdots, L\}$ evaluated at a frequency ω and defined as

$$\mathbf{C}(\omega) = \begin{bmatrix} C_{11}(\omega) & C_{12}(\omega) & \cdots & C_{1l}(\omega) & \cdots & C_{1L}(\omega) \\ C_{21}(\omega) & C_{22}(\omega) & \cdots & C_{2l}(\omega) & \cdots & C_{2L}(\omega) \\ \vdots & \vdots & \ddots & \vdots & \ddots & \vdots \\ C_{M1}(\omega) & C_{M2}(\omega) & \cdots & C_{Ml}(\omega) & \cdots & C_{ML}(\omega) \end{bmatrix}, \tag{3.3}$$

where $C_{ml}(\omega)$ is the electro-acoustic transfer function between the mth microphone R_m and the lth reproduction loudspeaker S_l calculated at ω. The above mentioned control task is achieved by introducing an $[L \times K]$ matrix of digital filters $\mathbf{W}(\omega)$ in the reproduction chain defined at a frequency ω as

$$\mathbf{W}(\omega) = \begin{bmatrix} W_{11}(\omega) & W_{12}(\omega) & \cdots & W_{1k}(\omega) & \cdots & W_{1K}(\omega) \\ W_{21}(\omega) & W_{22}(\omega) & \cdots & W_{2k}(\omega) & \cdots & W_{2K}(\omega) \\ \vdots & \vdots & \ddots & \vdots & \ddots & \vdots \\ W_{L1}(\omega) & W_{L2}(\omega) & \cdots & W_{Lk}(\omega) & \cdots & W_{LK}(\omega) \end{bmatrix}, \quad (3.4)$$

where $W_{lk}(\omega)$ is the filter driving the lth reproduction loudspeaker S_l and has $x_k(n)$ as its input. The filters $\mathbf{W}(\omega)$ are designed to produce the control signals $\underline{\mathbf{y}}(n)$ to drive the L reproduction loudspeakers such that the resulting sound field at the microphones $\hat{\underline{\mathbf{d}}}(n)$ is as close as possible to the desired sound field $\underline{\mathbf{d}}(n)$. The vectors $\underline{\mathbf{y}}(n)$ and $\hat{\underline{\mathbf{d}}}(n)$ are defined as

$$\underline{\mathbf{y}}(n) = \begin{bmatrix} y_1(n) & y_2(n) & \cdots & y_L(n) \end{bmatrix}^T, \quad (3.5)$$

$$\hat{\underline{\mathbf{d}}}(n) = \begin{bmatrix} \hat{d}_1(n) & \hat{d}_2(n) & \cdots & \hat{d}_M(n) \end{bmatrix}^T. \quad (3.6)$$

The set of desired sound fields given by the vector $\underline{\mathbf{d}}(n)$ is usually well correlated with the set of K input signals $\underline{\mathbf{x}}(n)$. Therefore, $\underline{\mathbf{d}}(n)$ may be considered as the result of filtering $\underline{\mathbf{x}}(n)$ through an $[M \times K]$ matrix of transfer functions $\mathbf{H}(\omega)$ (not shown in Fig. 3.1). This filtering operation may be expressed in the frequency domain as

$$\underline{\mathbf{D}}(\omega) = \mathbf{H}(\omega)\,\underline{\mathbf{X}}(\omega), \quad (3.7)$$

where $\underline{\mathbf{D}}(\omega)$ and $\underline{\mathbf{X}}(\omega)$ are the DFT of the time history of $\underline{\mathbf{d}}(n)$ and $\underline{\mathbf{x}}(n)$, respectively, as mentioned in Section 1.5. The matrix $\mathbf{H}(\omega)$ contains the DFT of the impulse responses $\{\underline{\mathbf{h}}_{mk} : m = 1, 2, \cdots, M,\ k = 1, 2, \cdots, K\}$ between the microphones and the audio signals evaluated at a frequency ω and defined as

$$\mathbf{H}(\omega) = \begin{bmatrix} H_{11}(\omega) & H_{12}(\omega) & \cdots & H_{1k}(\omega) & \cdots & H_{1K}(\omega) \\ H_{21}(\omega) & H_{22}(\omega) & \cdots & H_{2k}(\omega) & \cdots & H_{2K}(\omega) \\ \vdots & \vdots & \ddots & \vdots & \ddots & \vdots \\ H_{M1}(\omega) & H_{M2}(\omega) & \cdots & H_{Mk}(\omega) & \cdots & H_{MK}(\omega) \end{bmatrix}. \quad (3.8)$$

The key point in the above mentioned model is to design the matrix $\mathbf{W}(\omega)$ of $[L \times K]$ filters to achieve the control task. The design of those filters is discussed at length in this chapter. Examples illustrating the use of the generalised model in different audio reproduction applications are discussed in the next section.

3.1.2 Example Applications

In the generalised model introduced above, different audio reproduction applications may be described by defining different desired sets of sound fields $\underline{\mathbf{d}}(n)$ at the microphones. Consequently, it is the matrix $\mathbf{H}(\omega)$ that defines the nature of the application. Examples of the applications that may be described by the model are cross-talk cancellation, virtual source synthesis, concert hall simulation, correction of the responses of the reproduction loudspeakers, and active noise control.

In cross-talk cancellation (see Section 2.2.2), it is desired to exactly reproduce the input signals $\underline{\mathbf{x}}(n)$ at the microphones. In this case, $\underline{\mathbf{d}}(n) = \underline{\mathbf{x}}(n)$, and $\mathbf{H}(\omega) = \mathbf{I}$, the $[K \times K]$ identity matrix. Applying this for all frequencies, the transfer function between the mth microphone and the kth input signal has all its elements equal to unity, $\{\underline{\mathbf{H}}_{mk}(\omega) = \underline{\mathbf{1}} : m = k = 1, 2, \cdots, K\}$, which corresponds to a unit impulse response in the time domain $\underline{\mathbf{h}}_{mk}(n) = \delta(n)$[1]. Alternatively, K virtual sound images may be generated at the ears of $M/2$ (M even) listeners if $\underline{\mathbf{d}}(n)$ is the result of filtering the input signals $\underline{\mathbf{x}}(n)$ through the appropriate matrix of HRTFs as given by (2.10). The generalised model may also be used to simulate a concert hall by setting $\mathbf{H}(\omega)$ to be a matrix of transfer functions measured in a concert hall as mentioned in Section 2.1.6. Another application that may be described by the model is correcting the responses of the reproduction loudspeakers to obtain better sound quality. This may by achieved by letting $\mathbf{W}(\omega)$ be an inverse model of the loudspeakers' transfer functions. The most popular application of the generalised model is in active noise cancellation [56, 93, 109]. In such application, a set of K (undesired rather

[1]It will be shown in Section 3.3.7 that $\underline{\mathbf{h}}_{mk}(n) = \delta(n)$ results in a non-causal set of filters $\mathbf{W}(\omega)$. Causal filters produce a delayed version of the inputs $\hat{\underline{\mathbf{d}}}(n) = [x_1(n-\Delta_1)\ x_2(n-\Delta_2)\ \cdots\ x_K(n-\Delta_K)]$ at the microphones. In this case, the matrix $\mathbf{H}$ contains delayed unit impulse responses $\{\underline{\mathbf{h}}_{mk}(n) = \delta(n-\Delta_k) : m = k = 1, 2, \cdots, K\}$.

than desired) sound disturbances $\underline{\mathbf{d}}(n)$ are to be silenced at the microphones. This is achieved by filtering the input signals through $\mathbf{W}(\omega)$ to generate sound waves $\hat{\underline{\mathbf{d}}}(n)$ that are equal in amplitude but opposite in phase to the disturbances at the microphones' positions such that the net microphones' outputs $\underline{\mathbf{e}}(n)$ are minimised, where $\underline{\mathbf{e}}(n)$ is given by

$$\underline{\mathbf{e}}(n) = \left[\ e_1(n) \quad e_2(n) \quad \cdots \quad e_M(n)\ \right]^T. \tag{3.9}$$

The only difference between the above mentioned applications when using the generalised model is the desired response $\underline{\mathbf{d}}(n)$ at the microphones. Therefore, a generalised solution for $\mathbf{W}(\omega)$ that is valid for the whole class of applications may be obtained by solving the system equations for an arbitrary desired response. The solution for a specific application is then obtained by substituting its specific desired response into the general solution. The optimum solution $\mathbf{W}_{opt}(\omega)$ in the least mean square sense is derived in Section 3.2. An iterative approach to this optimum solution is discussed in Section 3.3.1. Examples of specific applications are discussed for active noise control, synthesis of virtual sound images, and cross-talk cancellation in Sections 3.3.5, 3.3.6, and 3.3.7, respectively.

3.2 The Optimum Least Mean Square Solution

The optimum Least Mean Square (LMS) solution for the system shown in Fig. 3.1 is obtained by minimising a performance index function $\xi(n)$, usually taken as the sum of the mean squared error signals

$$\xi(n) = \sum_{m=1}^{M} E\{\, e_m^2(n)\} = E\{\, \underline{\mathbf{e}}^T(n)\ \underline{\mathbf{e}}(n)\}, \tag{3.10}$$

where $E\{\cdot\}$ denotes the mathematical expectation. In the following discussion, all filters $\{\underline{\mathbf{w}}_{lk} \,:\, l = 1, 2, \cdots, L,\ k = 1, 2, \cdots, K\}$ are assumed to be Finite Impulse Response (FIR) digital filters, each of length N_w and defined as

$$\underline{\mathbf{w}}_{lk} = \left[\ w_{lk,0} \quad w_{lk,1} \quad \cdots \quad w_{lk,N_w-1}\ \right]^T. \tag{3.11}$$

The signal driving the lth reproduction loudspeaker $y_l(n)$ is the sum of the outputs of the filters $\{\underline{\mathbf{w}}_{lk} : k = 1, 2, \cdots, K\}$, which may be expressed as

$$y_l(n) = \underline{\mathbf{x}}_1^T(n)\, \underline{\mathbf{w}}_{l1} + \underline{\mathbf{x}}_2^T(n)\, \underline{\mathbf{w}}_{l2} + \cdots + \underline{\mathbf{x}}_K^T(n)\, \underline{\mathbf{w}}_{lK}, \qquad (3.12)$$

where the time history of the kth input signal $\underline{\mathbf{x}}_k(n)$ is defined as

$$\underline{\mathbf{x}}_k(n) = \begin{bmatrix} x_k(n) & x_k(n-1) & \cdots & x_k(n-N_w+1) \end{bmatrix}^T. \qquad (3.13)$$

Defining the $[KN_w \times 1]$ composite input signal vector $\underline{\mathbf{x}}(n)$ and the $[KN_w \times 1]$ composite weight vector $\underline{\mathbf{w}}_l$ as

$$\begin{aligned} \underline{\mathbf{x}}(n) &= \begin{bmatrix} \underline{\mathbf{x}}_1(n) & \underline{\mathbf{x}}_2(n) & \cdots & \underline{\mathbf{x}}_K(n) \end{bmatrix}^T, & (3.14)\\ \underline{\mathbf{w}}_l &= \begin{bmatrix} \underline{\mathbf{w}}_{l1} & \underline{\mathbf{w}}_{l2} & \cdots & \underline{\mathbf{w}}_{lK} \end{bmatrix}^T, & (3.15) \end{aligned}$$

equation (3.12) can be written as

$$y_l(n) = \underline{\mathbf{x}}^T(n)\; \underline{\mathbf{w}}_l. \qquad (3.16)$$

From (3.5) and (3.16), the $[L \times 1]$ vector of control signals $\underline{\mathbf{y}}(n)$ driving the reproduction loudspeakers is expressed as

$$\begin{bmatrix} y_1(n) \\ y_2(n) \\ \vdots \\ y_L(n) \end{bmatrix} = \begin{bmatrix} \underline{\mathbf{x}}(n) & \underline{\mathbf{0}} & \cdots & \underline{\mathbf{0}} \\ \underline{\mathbf{0}} & \underline{\mathbf{x}}(n) & \underline{\mathbf{0}} & \cdots \\ \vdots & \underline{\mathbf{0}} & \ddots & \underline{\mathbf{0}} \\ \underline{\mathbf{0}} & \cdots & \underline{\mathbf{0}} & \underline{\mathbf{x}}(n) \end{bmatrix}^T \begin{bmatrix} \underline{\mathbf{w}}_1 \\ \underline{\mathbf{w}}_2 \\ \vdots \\ \underline{\mathbf{w}}_L \end{bmatrix}. \qquad (3.17)$$

Further defining the $[KLN_w \times 1]$ composite vector $\underline{\mathbf{w}} = [\underline{\mathbf{w}}_1\ \underline{\mathbf{w}}_2\ \cdots\ \underline{\mathbf{w}}_L]^T$, and the $[KLN_w \times L]$ composite matrix $\mathbf{x}(n)$ of input signals given by the first factor in the right hand side of (3.17), (3.17) can be written compactly as

$$\underline{\mathbf{y}}(n) = \mathbf{x}^T(n)\, \underline{\mathbf{w}}. \qquad (3.18)$$

The system response vector $\hat{\underline{\mathbf{d}}}(n)$ results from filtering the input to the loudspeakers $\underline{\mathbf{y}}(n)$ through the matrix of electro-acoustic transfer functions $\mathbf{C}(\omega)$. The resulting component at the mth microphone is, therefore, given by

$$\hat{d}_m(n) = \underline{\mathbf{c}}_{m1} * \underline{\mathbf{y}}_1(n) + \underline{\mathbf{c}}_{m2} * \underline{\mathbf{y}}_2(n) + \cdots + \underline{\mathbf{c}}_{mL} * \underline{\mathbf{y}}_L(n), \qquad (3.19)$$

where all electro-acoustic impulse responses $\{\underline{\mathbf{c}}_{ml} : m = 1, 2, \cdots, M,\ l = 1, 2, \cdots, L\}$ are assumed to be FIR filters of length N_c defined as

$$\underline{\mathbf{c}}_{ml} = \left[\ c_{ml,0} \quad c_{ml,1} \quad \cdots \quad c_{ml,N_c-1}\ \right]^T. \tag{3.20}$$

Defining the time history of the signal driving the lth loudspeaker $y_l(n)$ as

$$\underline{\mathbf{y}}_l(n) = \left[\ y_l(n) \quad y_l(n-1) \quad \cdots \quad y_l(n-N_c+1)\ \right]^T, \tag{3.21}$$

and the $[M \times LN_c]$ composite matrix of acoustic impulse responses $\mathbf{c}$ as

$$\mathbf{c} = \begin{bmatrix} \underline{\mathbf{c}}_{11}^T & \underline{\mathbf{c}}_{12}^T & \cdots & \underline{\mathbf{c}}_{1L}^T \\ \underline{\mathbf{c}}_{21}^T & \underline{\mathbf{c}}_{22}^T & \cdots & \underline{\mathbf{c}}_{2L}^T \\ \vdots & \vdots & \ddots & \vdots \\ \underline{\mathbf{c}}_{M1}^T & \underline{\mathbf{c}}_{M2}^T & \cdots & \underline{\mathbf{c}}_{ML}^T \end{bmatrix}, \tag{3.22}$$

the $[M \times 1]$ vector $\hat{\underline{\mathbf{d}}}(n)$ of sound waves at the microphones can be expressed as

$$\hat{\underline{\mathbf{d}}}(n) = \mathbf{c} * \underline{\mathbf{y}}(n) = \mathbf{c} * [\mathbf{x}^T(n)\ \underline{\mathbf{w}}] = \mathbf{x}_f(n)\ \underline{\mathbf{w}}, \tag{3.23}$$

where the $[M \times KLN_w]$ matrix $\mathbf{x}_f(n)$ is given by

$$\mathbf{x}_f(n) = \begin{bmatrix} \underline{\mathbf{x}}_{f_{111}}(n) & \cdots & \underline{\mathbf{x}}_{f_{11K}}(n) & \cdots & \underline{\mathbf{x}}_{f_{1L1}}(n) & \cdots & \underline{\mathbf{x}}_{f_{1LK}}(n) \\ \underline{\mathbf{x}}_{f_{211}}(n) & \cdots & \underline{\mathbf{x}}_{f_{21K}}(n) & \cdots & \underline{\mathbf{x}}_{f_{2L1}}(n) & \cdots & \underline{\mathbf{x}}_{f_{2LK}}(n) \\ \vdots & \ddots & \vdots & \ddots & \vdots & \ddots & \vdots \\ \underline{\mathbf{x}}_{f_{M11}}(n) & \cdots & \underline{\mathbf{x}}_{f_{M1K}}(n) & \cdots & \underline{\mathbf{x}}_{f_{ML1}}(n) & \cdots & \underline{\mathbf{x}}_{f_{MLK}}(n) \end{bmatrix}, \tag{3.24}$$

where $\underline{\mathbf{x}}_{f_{mlk}}(n) = \underline{\mathbf{c}}_{ml} * \underline{\mathbf{x}}_k(n)$ is the last N_w samples of the result of filtering the kth input signal through the electro-acoustic impulse response between the mth microphone and the lth loudspeaker

$$\underline{\mathbf{x}}_{f_{mlk}}(n) = \left[\ x_{f_{mlk}}(n) \quad x_{f_{mlk}}(n-1) \quad \cdots \quad x_{f_{mlk}}(n-N_w+1)\ \right]^T. \tag{3.25}$$

The error vector $\underline{\mathbf{e}}(n)$ in Fig. 3.1 can then be expressed as

$$\underline{\mathbf{e}}(n) = \underline{\mathbf{d}}(n) - \mathbf{x}_f(n)\ \underline{\mathbf{w}}. \tag{3.26}$$

Substituting in the performance index (3.10) gives

$$\xi(n) = E\{\,\underline{\mathbf{d}}^T(n)\,\underline{\mathbf{d}}(n) - 2\,\underline{\mathbf{w}}^T\,\mathbf{x}_f^T(n)\,\underline{\mathbf{d}}(n) + \underline{\mathbf{w}}^T\,\mathbf{x}_f^T(n)\,\mathbf{x}_f(n)\,\underline{\mathbf{w}}\,\}. \tag{3.27}$$

The optimum LMS solution $\underline{\mathbf{w}}_{opt}(n)$ for the composite weight vector is the vector that minimises the performance index $\xi(n)$ given by (3.27). The optimum solution is, therefore, obtained by setting the first derivative (gradient) of $\xi(n)$ with respect to the composite weight vector $\underline{\mathbf{w}}$ to zero. The gradient is readily obtained from (3.27) to be

$$\underline{\nabla}(n) = \frac{\partial \xi(n)}{\partial \underline{\mathbf{w}}} = 2\,E\{\,\mathbf{x}_f^T(n)\,\mathbf{x}_f(n)\,\underline{\mathbf{w}} - \mathbf{x}_f^T(n)\,\underline{\mathbf{d}}(n)\,\}. \tag{3.28}$$

The optimum LMS solution $\underline{\mathbf{w}}_{opt}(n)$ is then obtained by setting $\underline{\nabla}(n)$ to zero

$$\underline{\mathbf{w}}_{opt}(n) = E\{\,(\mathbf{x}_f^T(n)\,\mathbf{x}_f(n))^{-1}\,\}\;E\{\,\mathbf{x}_f^T(n)\,\underline{\mathbf{d}}(n)\,\}, \tag{3.29}$$

and the corresponding minimum value of $\xi(n)$ is given by

$$\xi_{min} = E\{\,\underline{\mathbf{d}}^T(n)\underline{\mathbf{d}}(n)\}\; - \;E\{\,\underline{\mathbf{w}}_{opt}^T(n)\,\mathbf{x}_f^T(n)\,\underline{\mathbf{d}}(n)\,\}. \tag{3.30}$$

From (3.29), the optimum weight vector $\underline{\mathbf{w}}_{opt}(n)$ exists only if the matrix $(\mathbf{x}_f^T(n)\,\mathbf{x}_f(n))$ is non-singular. Since the matrix $\mathbf{x}_f(n)$ is the convolution between the input signals and the electro-acoustic transfer functions as given by (3.24), not only $\underline{\mathbf{x}}(n)$, but also $\mathbf{C}(\omega)$ influence the solution. Depending on the dimensions of $\mathbf{C}(\omega)$, three cases may be recognised:

1. **The number of loudspeakers equals the number of microphones ($L = M$):** In this case, $\mathbf{C}(\omega)$ is a square matrix. Consider $K = 1$ for simplicity and using frequency domain representations, the system of equations

   $$\underline{\mathbf{E}}(\omega) = \underline{\mathbf{D}}(\omega) - \mathbf{C}(\omega)\underline{\mathbf{Y}}(\omega) \tag{3.31}$$

   is fullydetermined. Provided that $\mathbf{C}(\omega)$ is non-singular, a unique solution for the control vector $\underline{\mathbf{Y}}(\omega)$ exists, which drives the error vector exactly to $\underline{\mathbf{0}}$, namely $\underline{\mathbf{Y}}_{opt}(\omega) = \mathbf{C}^{-1}(\omega)\,\underline{\mathbf{D}}(\omega)$.

2. **The number of loudspeakers is less than the number of microphones ($L < M$):** The matrix $\mathbf{C}(\omega)$ has more rows than columns and the system of equations (3.31) is overdetermined. There are more equations to solve than there are unknowns. In this case, provided that $(\mathbf{x}_f^T(n)\,\mathbf{x}_f(n))$ is positive definite, a unique global solution exists and is given by (3.29).

3. **The number of loudspeakers is greater than the number of microphones** ($L > M$)**:** There are less equations to solve than there are unknowns. The system of equations (3.31) is underdetermined and there exists an infinite number of solutions corresponding to infinite local minima on the error surface. In this case, extra constraints must be imposed to select one of those local solutions. A practical constraint may be limiting the power of the signals $\underline{\mathbf{y}}(n)$ driving the loudspeakers to avoid nonlinear distortion that may occur due to overloading the loudspeakers. Equivalently, the same result may be achieved by imposing the constraint on the values of the coefficients of $\underline{\mathbf{w}}$. The performance index in this case becomes

$$\xi(n) = E\{\,\underline{\mathbf{e}}^T(n)\,\mathbf{\Gamma}_e\,\underline{\mathbf{e}}(n) + \underline{\mathbf{w}}^T\,\mathbf{\Gamma}_w\,\underline{\mathbf{w}}\,\}, \qquad (3.32)$$

where $\mathbf{\Gamma}_e$ and $\mathbf{\Gamma}_w$ are (often diagonal) weighting matrixes [59, 110]. The optimum weight vector in this case is given by [59, 90, 93, 110]

$$\underline{\mathbf{w}}_{opt}(n) = E\{\,(\mathbf{x}_f^T(n)\,\mathbf{\Gamma}_e\,\mathbf{x}_f(n) + \mathbf{\Gamma}_w)^{-1}\,\}\;E\{\,\mathbf{x}_f^T(n)\,\mathbf{\Gamma}_e\,\underline{\mathbf{d}}(n)\,\}, \qquad (3.33)$$

and the corresponding minimum value of $\xi(n)$ is given by

$$\xi_{min} = E\{\,\underline{\mathbf{d}}^T(n)\mathbf{\Gamma}_e\,\underline{\mathbf{d}}(n)\} \;-\; E\{\,\underline{\mathbf{w}}_{opt}^T(n)\,\mathbf{\Gamma}_e\,\mathbf{x}_f^T(n)\,\underline{\mathbf{d}}(n)\,\}. \qquad (3.34)$$

The weighting $\mathbf{\Gamma}_w$ has also proven to improve the stability of the solution in the fullydetermined and overdetermined cases [59, 90]. Alternatively, the constraint may be imposed on the number of filter taps N_w, which enables an exact solution rather than a least mean square one [102, 103]. This latter option will be discussed in detail in Section 4.4

3.3 Iterative LMS Solutions

3.3.1 The Multiple Error Filtered-X LMS Algorithm

An alternative approach to obtaining the optimum LMS solution, other than direct calculation using (3.29) or (3.33), is to use an iterative algorithm such as the Multiple Error Least Mean Square (MELMS) algorithm [56, 58, 59, 108, 109, 110]. Such an adaptive approach is preferred over the direct calculation of $\underline{\mathbf{w}}_{opt}(n)$ since it offers *in-situ* design of the filters. It also enables a convenient method to readjust the filters whenever a change occurs in the electro-acoustic transfer functions as will be discussed in Chapter 4. The MELMS algorithm employs the steepest descent approach to search for the minimum of the performance index (3.10). This is achieved by successively updating the filters' coefficients by an amount proportional to the negative of the gradient $\underline{\nabla}(n)$,

$$\underline{\mathbf{w}}(n+1) = \underline{\mathbf{w}}(n) + \mu\,(-\,\underline{\nabla}(n)), \tag{3.35}$$

where μ is the step size that controls the convergence speed and the final misadjustment [144]. An approximation often used in such iterative LMS algorithms is to update the vector $\underline{\mathbf{w}}$ using the instantaneous value of the gradient $\tilde{\underline{\nabla}}(n)$ instead of its expected value $\underline{\nabla}(n) = E\{\tilde{\underline{\nabla}}(n)\}$ [144], leading to the well known LMS algorithm. Using (3.26) in (3.28), the gradient can be written as $\underline{\nabla}(n) = 2\,E\{\,-\mathbf{x}_f^T(n)\;\underline{\mathbf{e}}(n)\}$. The update equation for the MELMS algorithm is then given by replacing $\underline{\nabla}(n)$ in (3.35) by its instantaneous value,

$$\underline{\mathbf{w}}(n+1) = \underline{\mathbf{w}}(n) + 2\,\mu\,\mathbf{x}_f^T(n)\;\underline{\mathbf{e}}(n). \tag{3.36}$$

This update algorithm is often referred to as the Multiple Error Filtered-X (MEFX) algorithm. Implementation of (3.36) requires calculating the matrix $\mathbf{x}_f(n)$ given by (3.24), which implies measuring all the electro-acoustical transfer functions $\underline{\mathbf{c}}_{ml}$ and filtering each input signal through all ML transfer functions to construct the KLM elements of $\mathbf{x}_f(n)$. This is shown in Fig. 3.2, where the measured matrix of electro-acoustic transfer functions is represented by the block $\hat{\mathbf{C}}$ to distinguish it from the physical one represented by the block $\mathbf{C}$. The MEFX algorithm is known to be robust to estimation errors. It shows stable convergence properties as long as the phase error in any of the measured transfer functions at any frequency is less than $\pm 90°$ [56, 59, 105, 129, 144], while amplitude errors result in less accurate solutions [124]. On-line measurement of the matrix of electro-acoustic transfer functions $\hat{\mathbf{C}}$ is discussed in Section 4.7.

Similarly, the noisy steepest descent method may be used to iteratively approach the optimum solution in the case of weighted performance index. In this case

$$\underline{\nabla}(n) = 2\,E\{\,-\mathbf{x}_f^T(n)\,\mathbf{\Gamma}_e\,\underline{\mathbf{e}}(n)\;+\;\mathbf{\Gamma}_w\,\underline{\mathbf{w}}\}, \tag{3.37}$$

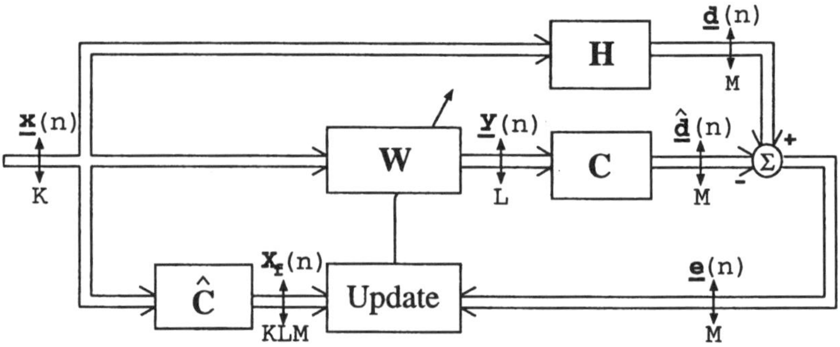

Figure 3.2: Block diagram of the multiple error filtered-x LMS algorithm.

which leads to the following update equation

$$\underline{\mathbf{w}}(n+1) = \underline{\mathbf{w}}(n) - \mu\,\mathbf{\Gamma}_w\,\underline{\mathbf{w}}(n) + 2\,\mu\,\mathbf{x}_f^T(n)\,\mathbf{\Gamma}_e\,\underline{\mathbf{e}}(n). \tag{3.38}$$

For $\mathbf{\Gamma}_w = \text{diag}\{\gamma_1\ \gamma_2\ \cdots\ \gamma_{\text{KLN}_w}\}$, each filter weight w_i is independently weighted by the weighting factor γ_i. In this case, when $\mathbf{\Gamma}_e = \mathbf{I}$, and letting $\mathbf{\Gamma} = \mathbf{I} - \mu\ \mathbf{\Gamma}_w$, the update equation becomes the multiple error leaky LMS algorithm

$$\underline{\mathbf{w}}(n+1) = \mathbf{\Gamma}\,\underline{\mathbf{w}}(n) + 2\,\mu\,\mathbf{x}_f^T(n)\ \underline{\mathbf{e}}(n). \tag{3.39}$$

3.3.2 Convergence Properties of the MEFX Algorithm

The convergence properties of the regular LMS algorithm (when $\mathbf{C}(\omega) = \mathbf{I}$) for a single input are determined by the eigenvalues of the autocorrelation matrix of the input signal [144]. Since the input signals to the MEFX update algorithm are the filtered signals given by the elements of the matrix $\mathbf{x}_f(n)$, the convergence properties of the MEFX are expected to be determined by the matrix $\mathbf{R}_f(n) = E\{\mathbf{x}_f^T(n)\ \mathbf{x}_f(n)\}$ [105, 129]. The matrix $\mathbf{R}_f(n)$ is, in general, positive definite for $L \leq M$, and the error surface has a unique global minimum. In this section, the convergence properties of the time domain MEFX are discussed using frequency domain representations of signals [37, 38, 64]. Frequency domain analysis decouples the system to approximately uncorrelated frequency bins, and the convergence properties are approximately determined by independent modes at each frequency. Furthermore, frequency domain analysis allows the study of the effects of the electro-acoustic transfer functions on the system performance, which is discussed in detail in Section 3.3.3.

From the definition of the matrix $\mathbf{x}_f(n)$ given in (3.24), the frequency domain version $\mathbf{X}_f(n,\omega)$ at a frequency ω may be expressed as

$$\mathbf{X}_f(n,\omega) = \begin{bmatrix} C_{11}X_1 & \cdots & C_{11}X_K & \cdots & C_{1L}X_1 & \cdots & C_{1L}X_K \\ C_{21}X_1 & \cdots & C_{21}X_K & \cdots & C_{2L}X_1 & \cdots & C_{2L}X_K \\ \vdots & \ddots & \vdots & \ddots & \vdots & \ddots & \vdots \\ C_{M1}X_1 & \cdots & C_{M1}X_K & \cdots & C_{ML}X_1 & \cdots & C_{ML}X_K \end{bmatrix}, \tag{3.40}$$

where the explicit dependency on (n,ω) has been dropped inside the matrix for clarity. Representing the system at one frequency ω as in (3.40) reduces the dimensions of $\mathbf{X}_f(n,\omega)$ from $[M \times KLN_w]$ to $[M \times KL]$. Consequently, the dimensions of $\underline{\mathbf{W}}(n,\omega)$ reduces to $[KL \times 1]$ corresponding to KL filters each has a single coefficient at ω. However, it should be kept in mind that there are as many equations of the form (3.40) as there are frequencies.

Assuming slow adaptation, the adaptive filters may be considered time invariant linear filters during a short period of time. In this period, the MEFX algorithm can be approximated in the frequency domain using the following equations

$$\underline{\mathbf{E}}(n,\omega) = \underline{\mathbf{D}}(n,\omega) + \mathbf{X}_f(n,\omega)\,\underline{\mathbf{W}}(n,\omega), \tag{3.41}$$
$$\underline{\mathbf{W}}(n+1,\omega) = \underline{\mathbf{W}}(n,\omega) + 2\mu'\mathbf{X}_f^H(n,\omega)\,\underline{\mathbf{E}}(n,\omega), \tag{3.42}$$

where $\cdot^H$ denotes the complex conjugate transpose and μ' is the frequency domain step size. Upper-case letters are the DFT of the corresponding lower-case variables at iteration n as mentioned in Section 1.5. Substituting (3.41) into (3.42) and taking the mathematical expectation, (3.42) may be approximated by

$$E\{\underline{\mathbf{W}}(n+1,\omega)\} - \underline{\mathbf{W}}_{opt}(\omega) \approx (\mathbf{I} - 2\mu'\mathbf{R}_f(\omega))\,[E\{\underline{\mathbf{W}}(n,\omega)\} - \underline{\mathbf{W}}_{opt}(\omega)], \tag{3.43}$$

where $\mathbf{R}_f(\omega) = E\{\mathbf{X}_f^H(n,\omega)\,\mathbf{X}_f(n,\omega)\}$, and $\underline{\mathbf{W}}_{opt}(\omega)$ is the Fourier transform of the optimal weight vector $\underline{\mathbf{w}}_{opt}(n)$ given by (3.29).

Since $\mathbf{R}_f(\omega)$ is, in general, not diagonal, the components of $\underline{\mathbf{W}}(n,\omega)$ are cross-coupled. Provided that $\mathbf{R}_f(\omega)$ is symmetric and positive, it may be factored into its eigenvalues and eigenvectors [144],

$$\mathbf{R}_f(\omega) = \mathbf{Q}(\omega)\ \mathbf{\Lambda}_f(\omega)\ \mathbf{Q}^H(\omega), \tag{3.44}$$

where $\mathbf{\Lambda}_f(\omega) = \text{diag}\{\lambda_{f_1}(\omega)\ \lambda_{f_2}(\omega)\ \cdots\ \lambda_{f_{KL}}(\omega)\}$ is a diagonal matrix of the eigenvalues of $\mathbf{R}_f(\omega)$ and $\mathbf{Q}(\omega) = [\underline{\mathbf{Q}}_1(\omega)\ \underline{\mathbf{Q}}_2(\omega)\ \cdots\ \underline{\mathbf{Q}}_{KL}(\omega)]$ is the matrix of eigenvectors of $\mathbf{R}_f(\omega)$ satisfying $\mathbf{Q}\,\mathbf{Q}^H = \mathbf{Q}^H\,\mathbf{Q} = \mathbf{I}$. Substituting (3.44) in (3.43) and using the rotated weight difference vector $\underline{\mathbf{V}}(n,\omega) = \mathbf{Q}^H\,(E\{\underline{\mathbf{W}}(n,\omega)\} - \underline{\mathbf{W}}_{opt}(\omega))$ results in

$$\underline{\mathbf{V}}(n+1,\omega) = (\mathbf{I} - 2\mu'\,\mathbf{\Lambda}_f(\omega))\ \underline{\mathbf{V}}(n,\omega). \tag{3.45}$$

Equation (3.45) may be solved by induction to obtain $\underline{\mathbf{V}}(n,\omega)$ as a function of the initial rotated weight difference vector $\underline{\mathbf{V}}(0,\omega)$ giving

$$\underline{\mathbf{V}}(n,\omega) = (\mathbf{I} - 2\mu'\,\mathbf{\Lambda}_f(\omega))^n\ \underline{\mathbf{V}}(0,\omega). \tag{3.46}$$

Since $(\mathbf{I} - 2\mu'\,\mathbf{\Lambda}_f(\omega))$ is now diagonal, (3.46) can be decomposed into LM independent equations, each corresponding to a fundamental system mode at ω. From these decoupled equations, it can be seen that the MEFX is stable at ω if

$$|(1 - 2\mu'\,\lambda_{f_i}(\omega))| < 1, \qquad i = 1, 2, \cdots, KL. \tag{3.47}$$

This defines the limits for the step size to be $0 < \mu' < 1/\lambda_{f_{max}}(\omega)$, where $\lambda_{f_{max}}(\omega)$ is the largest eigenvalue of the matrix $\mathbf{R}_f(\omega)$ at ω. The algorithm is stable at all frequencies if μ' satisfies

$$0 < \mu' < \frac{1}{\max\limits_{\omega}\{\lambda_{f_{\max}}(\omega)\}}\,. \tag{3.48}$$

If the condition (3.48) is satisfied, then as $n \to \infty$, the rotated weight difference vector $\underline{\mathbf{V}}(n,\omega) \to \underline{\mathbf{0}}$, and (on average) the adaptive weights at ω approach their optimum solutions $E\{\underline{\mathbf{W}}(n,\omega)\} \to \underline{\mathbf{W}}_{opt}(\omega)$. However, each individual coefficient $W_i(n,\omega)$, $i = 1, 2, \cdots, KL$ approaches its optimal value following an approximate exponential decaying envelope with time constant [144]

$$\tau_i(\omega) \approx \frac{1}{2\mu'\lambda_{f_i}(\omega)}\,. \tag{3.49}$$

The convergence speed of the whole system is determined by the time constant of the slowest mode over the whole frequency range of interest, which is given by [38, 144],

$$\tau_{max} > \frac{\max_{\omega}\{\lambda_{f_{\max}}(\omega)\}}{2\min_{\omega}\{\lambda_{f_{\min}}(\omega)\}}. \tag{3.50}$$

Therefore, the convergence speed of the time domain adaptive weights is subject to the ratio of the maximum to minimum eigenvalues of the matrix $\mathbf{R}_f(\omega)$ over the whole frequency range of interest. Since both $\underline{\mathbf{X}}(\omega)$ and $\mathbf{C}(\omega)$ contribute to $\mathbf{R}_f(\omega)$, the convergence speed of the MEFX algorithm is affected by both the input signals' statistics and the matrix $\mathbf{C}(\omega)$. The dependency of the convergence speed of the conventional LMS algorithm on the statistics of the input signal is well known in traditional adaptive filters literature such as [144]. The effects of $\mathbf{C}(\omega)$ and their physical interpretations are discussed in the next section.

3.3.3 Effects of the Electro-Acoustic Transfer Functions

The frequency domain matrix $\mathbf{X}_f(n,\omega)$ defined in (3.40) may be expressed as

$$\mathbf{X}_f(n,\omega) = \mathbf{C}(\omega)\,\mathbf{X}(n,\omega), \tag{3.51}$$

where $\mathbf{C}(\omega)$ is the matrix of electro-acoustic transfer functions and $\mathbf{X}(n,\omega)$ is the Fourier transform of $\mathbf{x}^T(n)$ defined in (3.17), both evaluated at ω. To decouple the effects of $\mathbf{X}$ and $\mathbf{C}$ and study the effects of $\mathbf{C}(\omega)$ only, the input signals $\{x_k(n)\,:\,k = 1, 2, \cdots, K\}$ are assumed white noise with zero mean and variances σ_1^2, σ_2^2, $\cdots$, σ_K^2 that are uncorrelated with each other and with $\mathbf{C}(\omega)$. From (3.51), the matrix $\mathbf{R}_f(\omega)$ may be written as $\mathbf{R}_f(\omega) = E\{\mathbf{X}^H(n,\omega)\,\mathbf{C}^H(\omega)\,\mathbf{C}(\omega)\,\mathbf{X}(n,\omega)\}$. The determinant of $\mathbf{R}_f(\omega)$ can then be expressed as [38, 39]

$$\det\{\mathbf{R}_f(\omega)\} = |\mathbf{R}_f(\omega)| = N_B^K\,\sigma_1^2\,\sigma_2^2\,\cdots\,\sigma_K^2\,|\mathbf{C}^H(\omega)\,\mathbf{C}(\omega)|^K, \tag{3.52}$$

where N_B is the number of frequency domain filter coefficients (the size of the DFT). Since the determinant of a matrix is equal to the product of its eigenvalues, it is clear from (3.52) that the eigenvalues of $\mathbf{R}_f(\omega)$ are influenced by the eigenvalues of $\mathbf{C}^H(\omega)\,\mathbf{C}(\omega)$. Furthermore, the smallest eigenvalue of $\mathbf{R}_f(\omega)$ is determined only by the smallest eigenvalue of

$\mathbf{C}^H(\omega)\,\mathbf{C}(\omega)$ at ω, or equivalently by the determinant of $\mathbf{C}^H(\omega)\,\mathbf{C}(\omega)$ at that frequency. The slowest mode is, therefore, completely determined by the lowest value of $\det\{\mathbf{C}^H(\omega)\,\mathbf{C}(\omega)\}$ over the whole frequency range of interest.

To assist understanding the physical meaning of the above mentioned result, we consider the system $[K \times L \times M = 1 \times 1 \times 1]$. In this case, both $\mathbf{C}(\omega)$ and $\mathbf{R}_f(\omega)$ are $[1\times1]$ matrixes and there is only one eigenvalue $\lambda_f(\omega)$ at each frequency. This eigenvalue can be evaluated from $\lambda_f(\omega) = \det\{\mathbf{C}^H(\omega)\,\mathbf{C}(\omega)\} = |C(\omega)|^2$. From (3.47), and assuming white noise input of zero mean and variance σ^2, the stability condition becomes

$$0 < \mu' < \frac{1}{\max_{\omega}\{|C(\omega)|^2\}\; N_B\,\sigma^2}. \tag{3.53}$$

This shows that the step size μ' is limited by the maximum power gain of the single transfer function between the loudspeaker and the microphone. From (3.49), the time constant $\tau(\omega)$ becomes

$$\tau(\omega) \approx \frac{1}{2\,\mu'|C(\omega)|^2\, N_B\,\sigma^2}, \tag{3.54}$$

which shows clearly that the smallest power gain defines the slowest mode. In an extreme case, when $|C(\omega)| = 0$ at any frequency ω, the time constant at that frequency is infinitely long and the adaptive filter will never reach its optimal solution.

In multichannel systems, the influence of $\mathbf{C}(\omega)$ is enhanced due to the interaction between the different electro-acoustic transfer functions comprising $\mathbf{C}(\omega)$. To show this, consider the system $[K\times L\times M = 2\times2\times2]$, where

$$\det\{\mathbf{C}^H(\omega)\,\mathbf{C}(\omega)\} = (C_{11}C_{22} - C_{21}C_{12})\,(C_{11}^*C_{22}^* - C_{21}^*C_{12}^*). \tag{3.55}$$

As mentioned above, whenever the determinant in (3.55) equals zero (or a very small value) at any frequency, the whole adaptive process is slowed. From (3.55), the following three cases which result in $\det\{\mathbf{C}^H\,\mathbf{C}\} \to 0$ may be recognised:

1. $C_{11}(\omega) = C_{12}(\omega) = 0$ or $C_{22}(\omega) = C_{21}(\omega) = 0$. This occurs when all transfer functions from all loudspeakers to any of the microphones are zeros at the same frequency, leading to a zero row in

the matrix $\mathbf{C}(\omega)$. This effectively reduces the number of microphones at this frequency and reduces the rank of the matrix $\mathbf{C}(\omega)$ at that specific frequency. This situation may occur in practice when the frequency responses of all loudspeakers have a notch at the same frequency, or the concerned microphone is insensitive at a given frequency.

2. $C_{11}(\omega) = C_{21}(\omega) = 0$ or $C_{22}(\omega) = C_{12}(\omega) = 0$. This occurs when all transfer functions from any of the loudspeakers to all microphones are zeros at the same frequency leading to a zero column in the matrix $\mathbf{C}(\omega)$. In such a situation, the related loudspeaker is not effective at that frequency, which also reduces the rank of $\mathbf{C}$ by one at that specific frequency. Similar practical causes as those mentioned above may also be valid in this case.

3. $C_{11}(\omega)\, C_{22}(\omega) = C_{21}(\omega)\, C_{12}(\omega)$. This occurs in special symmetrical acoustic arrangements. If the distances from two or more loudspeakers to each microphone are equal, two or more linearly dependent columns in the matrix $\mathbf{C}(\omega)$ are created. Similarly, if the distances from two or more microphones to each loudspeaker are equal, two linearly dependent rows are created.

Since a minimum 3D sound system consists of two loudspeakers (stereophonic set-up) and two microphones (two ears), care must be taken in choosing the positions of the loudspeakers relative to the microphones so that none of the above mentioned ill-conditions occurs. Improving the convergence speed of the adaptive filters by decorrelating the effect of $\mathbf{C}(\omega)$ is discussed in Section 4.8, while the effects of an ill-conditioned $\mathbf{C}(\omega)$ and the choice of the positions of the loudspeakers are discussed in detail in Section 4.5.

3.3.4 The Causality Constraint

Reducing the error signals $\underline{\mathbf{e}}(n)$ in Fig. 3.2 to zeros makes the actual system outputs $\hat{\underline{\mathbf{d}}}(n)$ equal to the desired responses $\underline{\mathbf{d}}(n)$. Since the inputs $\underline{\mathbf{x}}(n)$ to the cascade combination of $\mathbf{W}(\omega)$ and $\mathbf{C}(\omega)$ are the same as the inputs to the system $\mathbf{H}(\omega)$, the control problem may be considered as adjusting $\mathbf{W}(\omega)$ so that

$$\mathbf{C}(\omega)\, \mathbf{W}(\omega) = \mathbf{H}(\omega), \tag{3.56}$$

where the dimensions of $\mathbf{C}(\omega)$, $\mathbf{W}(\omega)$, and $\mathbf{H}(\omega)$ are $[M \times L]$, $[L \times K]$ and $[M \times K]$, respectively. This shows that $\mathbf{W}(\omega)$ is a combination of forward modelling of $\mathbf{H}(\omega)$ and inverse modelling of $\mathbf{C}(\omega)$. The success of this modelling problem is, in general, limited by the following factors:

- Being physical systems, the elements of $\mathbf{C}(\omega)$ correspond to causal impulse responses. The input signals to the loudspeakers $\underline{\mathbf{y}}(n)$ are delayed as they go through the physical system $\mathbf{C}(\omega)$. Assuming for the moment that $\mathbf{H}(\omega)$ does not introduce any delay, the desired responses $\underline{\mathbf{d}}(n)$ arrive at the microphones before $\hat{\underline{\mathbf{d}}}(n)$. This requires the filters $\mathbf{W}(\omega)$ to be predictors, a task that can only be performed approximately by a causal filter in a statistical sense [144]. To avoid this prediction solution, $\mathbf{H}(\omega)$ must have delay components that are longer than the delays introduced by $\mathbf{C}(\omega)$.

- The elements of $\mathbf{C}(\omega)$ are often non-minimum phase impulse responses and their transfer functions have zeros outside the unit circle in the z-plane. Inverting such a non-minimum phase response results in poles outside the unit circle. Such an inverse is stable only if the impulse response is left-handed in time (non-causal). Since a delayed non-causal impulse response may be approximated by a causal impulse response truncated in time [144], a delayed version of the non-causal solution may be approximated by $\mathbf{W}(\omega)$. This delay may be introduced by delaying the desired response $\underline{\mathbf{d}}(n)$.

The above discussion suggests that for successful adaptation of $\mathbf{W}(\omega)$, it is essential to analyse the delay introduced by the physical transfer functions between the loudspeakers and the microphones. When the desired responses $\underline{\mathbf{d}}(n)$ reach the microphones before $\hat{\underline{\mathbf{d}}}(n)$, the optimum solutions for the filters $\mathbf{W}(\omega)$ are non-causal. In most audio reproduction applications it is possible to add artificial delay to $\underline{\mathbf{d}}(n)$ using the matrix $\mathbf{H}(\omega)$. This shifts the required non-causal solutions to the right along the time axis, which allows the FIR filters $\mathbf{W}$ to approximate a truncated versions of the non-causal solutions. In active noise cancellation (Section 3.3.5), this delay may be achieved by placing all secondary sources closer to the control sensors than the sources of the disturbance. In cross-talk cancellation, delay components have to be added in $\mathbf{H}(\omega)$ as mentioned in Section 3.3.7, and the reproduced audio signals are delayed versions

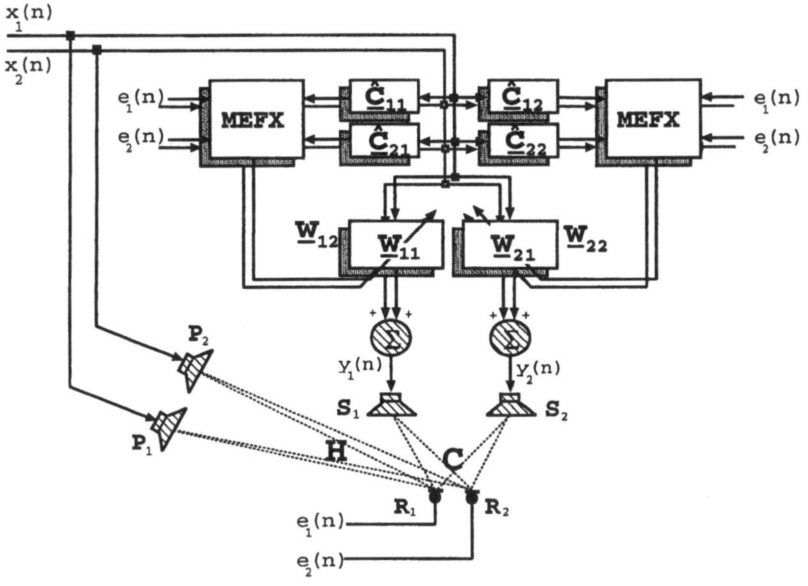

Figure 3.3: Block diagram of the MEFX algorithm for a $[K \times L \times M = 2 \times 2 \times 2]$ system in an active noise control setting.

of the recorded input signals. In the following, it is assumed that sufficient delay components are introduced whenever necessary such that the adaptive filters are good approximations of the optimum solutions.

3.3.5 Adaptive Active Noise Cancellation

The detailed implementation of the MEFX algorithm is further explained by unpacking the composite vector $\underline{\mathbf{w}}(n)$ in (3.36) into its individual filters $\{\underline{\mathbf{w}}_{lk}(n) \,:\, l = 1, 2, \cdots, L,\; k = 1, 2, \cdots, K\}$ giving

$$\underline{\mathbf{w}}_{lk}(n+1) = \underline{\mathbf{w}}_{lk}(n) + 2\,\mu \sum_{m=1}^{M} e_m(n)\; \underline{\mathbf{x}}_{f_{lmk}}(n), \qquad (3.57)$$

where $\underline{\mathbf{x}}_{f_{lmk}}(n)$ is the result of filtering the kth input through the measured electro-acoustic impulse response $\hat{\underline{\mathbf{C}}}_{ml}$ between the mth microphone and the lth loudspeaker. The detailed implementation of (3.57) is visually illustrated in Fig. 3.3 for a $[K \times L \times M = 2 \times 2 \times 2]$ active noise control system. In this system, it is desired to reduce the sound field due to the primary sources P_1 and P_2 at two receiver points (microphones) R_1 and R_2 using an anti-sound field generated by the secondary loudspeakers S_1 and S_2. Using frequency domain representations, the inputs to the secondary sources $Y_1(\omega)$ and $Y_2(\omega)$ are controlled by four adaptive filters $\{\underline{\mathbf{W}}_{lk}(\omega) \,:\, l = 1, 2,\ k = 1, 2\}$ to achieve the above cancellation task. The desired response is the sound field generated by the primary sources P_1 and P_2 at the microphones' positions: $\underline{\mathbf{D}}(\omega) = [D_1(\omega)\ D_2(\omega)] = \mathbf{H}(\omega)\ \underline{\mathbf{X}}(\omega)$, where $\mathbf{H}(\omega)$ is the matrix of electro-acoustic transfer functions between the two microphones and the primary sources and $\underline{\mathbf{X}}(\omega) = [X_1(\omega)\ X_2(\omega)]$. The microphones' outputs are, therefore, the *sum* of the sound fields due to the primary and the secondary sources $\underline{\mathbf{E}}(\omega) = [E_1(\omega)\ E_2(\omega)] = \underline{\mathbf{D}}(\omega) + \hat{\underline{\mathbf{D}}}(\omega)$, where $\hat{\underline{\mathbf{D}}}(\omega) = [\hat{D}_1(\omega)\ \hat{D}_2(\omega)] = \mathbf{C}(\omega)\ \underline{\mathbf{Y}}(\omega)$ is the (unmeasurable) sound field due to the secondary sources alone at the microphones.

At each time sample, the update algorithm adjusts the coefficients of the adaptive filters to minimise the error signals measured by the microphones using the MEFX algorithm (3.57)[2]. Four update blocks are needed, one for each adaptive filter. Those are indicated by the MEFX boxes in Fig. 3.3. According to (3.57), each of the update blocks requires two filtered input signals, in total $KLM = 8$ filtered input signals are needed. The filtered input signals are calculated by filtering each of the two input signals through the four measured impulse responses $\{\hat{\underline{\mathbf{C}}}_{ml}(\omega) \,:\, m = 1, 2,\ l = 1, 2\}$ that are estimates of the physical electro-acoustic transfer functions between the secondary sources and the microphones represented in Fig. 3.3 by the matrix $\mathbf{C}$.

After successful convergence, the sound waves generated by S_1 and S_2 at R_1 and R_2 are equal in magnitude and opposite in phase to that generated by P_1 and P_2, and reduction in the net sound field at R_1 and R_2 results. This may be expressed mathematically as $\mathbf{H}(\omega)\,\underline{\mathbf{X}}(\omega) = -\mathbf{C}(\omega)\,\mathbf{W}(\omega)\,\underline{\mathbf{X}}(\omega)$, and the solution to the matrix of control filters is given by

$$\mathbf{W}(\omega) = -\,\mathbf{C}^{-1}(\omega)\,\mathbf{H}(\omega). \tag{3.58}$$

[2]Since (3.57) was derived assuming that the error is formed by the difference between rather than the sum of $\underline{\mathbf{D}}(\omega)$ and $\hat{\underline{\mathbf{D}}}(\omega)$, the + sign in (3.57) must be changed to − sign.

3.3.6 Adaptive Virtual Sound Sources

A comparison between (3.58) and (2.15) shows that the system in Fig. 3.3 is exactly (except for a – sign) what is needed to generate two virtual sound images at the positions of the primary sources P_1 and P_2 at the ears of one listener when the two microphones are positioned inside the listener's ear canals. By controlling the sound at $M = 2B$ microphones, the same sound images are perceived by B listeners, and by increasing the number of inputs, more images are created. This suggests a procedure for implementing the loudspeaker display system discussed in Section 2.2.2. With probe microphones inserted in the listener's ear canals, the set of adaptive filters are adjusted to cancel uncorrelated white noise signals from physical loudspeakers placed where virtual sound images are required[3]. After successful conversion, the physical primary sources are disconnected and the coefficients of the filters are multiplied by -1 to compensate for the sign difference mentioned above. Monophonic signals filtered through the previously obtained filters will move the auditory events to the positions where the primary sources have been. This procedure has been successfully used in [2, 3, 83, 100, 131] and proved to have the following advantages compared to other methods:

- The listener's own HRTFs are used to design the filters, therefore, correct spectral cues are maintained.
- The solution also includes the room impulse response, therefore, maintains the distance cues and the environmental context. This solves the In-Head-Localisation (IHL) problem and eliminates the need for artificial reverberation.
- Direct inversion of the electro-acoustic transfer function matrix $\mathbf{C}(\omega)$ that is required for the cross-talk cancellation is avoided by iteratively searching the optimum solution in the least mean squared sense.
- The binaural synthesis (convolution with the HRTF matrix $\mathbf{H}(\omega)$) and the cross-talk cancellation (inversion of $\mathbf{C}(\omega)$) are combined. This is both more efficient and numerically more stable.

[3]The adaptation process requires first measuring all electro-acoustic transfer functions $\underline{\hat{\mathbf{C}}}_{ml}$. On-line estimation of those transfer functions is discussed in detail in Section 4.7

- The problem is mapped from the difficult domain of virtual sound image synthesis to the well-developed one of multichannel active noise control (ANC). This not only allows employing the techniques used in ANC systems, but also facilitates describing the system performance in terms of the sound attenuation achieved at the listener's eardrums. The larger this attenuation, the more realistic the virtual source is perceived.

- Active noise control systems are real-time systems, therefore, the above procedure allows real-time design and implementation of the system's filters. This differs from the commonly used methods of measuring and storing a set of HRTFs, calculating the cross-talk cancellation filters, and real-time filtering the audio signals through those previously calculated fixed filters.

In spite of the above mentioned advantages, adaptive filters approach to virtual source synthesis suffers from the following drawbacks:

- Including the room impulse response in the filters limits the virtual space to the same measurement environment.

- The listener is asked to insert a pair of microphones inside his/her ears, which may be objectionable. However, this is the only approach to obtaining individualised HRTFs.

- White noise or chirp signals that are spectrally rich must be used in the identification and adaptation stages[4].

- The electro-acoustic transfer functions $\mathbf{C}(\omega)$ and $\mathbf{H}(\omega)$ are very complex functions of frequency and space coordinates. The impulse response of a room may also last for several hundreds of milliseconds. At a sampling frequency of 44.1 kHz, thousands of FIR coefficients are required to properly model these transfer functions. The adaptive filters are, therefore, of high order, which makes system implementation in real-time a great challenge. Efficient implementation of the filtering and adaptation operations are, therefore, essential[5].

---

[4] Adaptation using audio signals are also considered in Section 4.7.

[5] Efficient implementations of adaptive filters are discussed in detail in Section 3.4.

- At high frequencies, the acoustic wavelength is very small. Therefore, the solution obtained by the adaptive process is valid only in a very small area in space. Therefore, the listener's head must be fixed during the adaptation and filtering operations[6].

3.3.7 Adaptive Cross-Talk Cancellation

The cross-talk cancellation mentioned in Section 2.2.2 may also be realised using adaptive filters. The system shown in Fig. 3.3 can be used for this purpose after a few modifications. In cross-talk cancellation, it is desired to reproduce $x_1(n)$ at R_1 and $x_2(n)$ at R_2. The desired response is, therefore, given by $\underline{\mathbf{d}}(n) = \underline{\mathbf{x}}(n)$, and the adaptive filters are required to model the inverse of the matrix $\mathbf{C}(\omega)$. For the inverse to be realisable using FIR filters, delayed input signals are used to calculate the desired responses as mentioned in Section 3.3.4,

$$\begin{bmatrix} d_1(n) \\ d_2(n) \end{bmatrix} = \begin{bmatrix} x_1(n-\Delta_1) \\ x_2(n-\Delta_2) \end{bmatrix}, \tag{3.59}$$

where Δ_1 and Δ_2 are delays that are assumed longer than that introduced by the electro-acoustic transfer functions comprising $\mathbf{C}(\omega)$. Since the microphones' outputs in this application are due to S_1 and S_2 only, they correspond to the vector $\hat{\underline{\mathbf{d}}}(n)$ in Fig. 3.1. The error signals, to be minimised by the adaptive algorithm, are then constructed by *electrically* subtracting $\hat{\underline{\mathbf{d}}}(n)$ from $\underline{\mathbf{d}}(n)$

$$\begin{bmatrix} e_1(n) \\ e_2(n) \end{bmatrix} = \begin{bmatrix} x_1(n-\Delta_1) - \hat{d}_1(n) \\ x_2(n-\Delta_2) - \hat{d}_2(n) \end{bmatrix}. \tag{3.60}$$

Provided that the FIR filters $\mathbf{W}$ are of sufficiently high order to accommodate the inverse solution, the mean square error is minimised by the adaptive algorithm. After conversion, the sound pressure at the microphones R_1 and R_2 are the best least squares estimations of $x_1(n-\Delta_1)$ and $x_2(n-\Delta_2)$, respectively.

[6]This problem is common to all 3D sound systems and will be treated separately in Chapter 4.

3.4 Efficient Implementations

In reverberant acoustic environments, the impulse response between two points may last for several hundreds of milliseconds. At the standard audio compact disc sampling frequency of 44.1 kHz, thousands of FIR coefficients are needed to properly model and store such an impulse response. Adaptive 3D sound systems require estimating, filtering through, and updating the coefficients of many of these large filters. This implies a huge amount of computations, which makes real-time implementation on reasonable hardware resources a difficult task. Reducing the system complexity is, therefore, essential for real-time implementation. In this section, two approaches are introduced to decrease the number of required calculations. The adjoint LMS algorithm discussed in Section 3.4.1 uses filtered error signals rather than filtered input signals in the update process. This leads to a huge computational saving as the system dimensions increase. Significant saving may further be achieved by implementing the convolution and correlation operations in the frequency domain using block processing techniques. This leads to the Block Frequency Domain Adaptive Filters (BFDAF) discussed in Section 3.4.2.

3.4.1 The Adjoint LMS Algorithm

Setting the electro-acoustic transfer function matrix $\mathbf{C}(\omega)$ in Fig. 3.2 at each frequency to be the identity matrix $\mathbf{I}$ results in a regular adaptive filter system [144]. In this case, $\underline{\hat{\mathbf{d}}}(n) = \underline{\mathbf{y}}(n)$ and the adaptive filters' outputs are directly observable from the error vector $\underline{\mathbf{e}}(n) = \underline{\mathbf{d}}(n) - \underline{\mathbf{y}}(n)$. The steepest descent update given by (3.36) becomes

$$\underline{\mathbf{w}}(n+1) = \underline{\mathbf{w}}(n) + 2\,\mu\,\mathbf{x}^T(n)\;\underline{\mathbf{e}}(n), \qquad (3.61)$$

where $\mathbf{x}(n)$ is defined in (3.17). The adaptive process in (3.61) basically aims at minimising the crosscorrelation between the input signals $\mathbf{x}(n)$ and the error signals $\underline{\mathbf{e}}(n)$. When this crosscorrelation reaches its minimum value, the filters are as close as possible to their optimal solutions. For correct adaptation, it is essential that the crosscorrelation is performed between correctly aligned history samples of $\underline{\mathbf{e}}(n)$ and $\mathbf{x}(n)$.

Introducing the matrix $\mathbf{C}(\omega)$ into the system results in $\underline{\hat{\mathbf{d}}}(n)$ being a filtered version of $\underline{\mathbf{y}}(n)$. This has two severe consequences on the adaptation process. The first is that the crosscorrelation is performed between two misaligned time sequences, since $\underline{\mathbf{y}}(n)$ is delayed while propagating through the matrix of acoustical transfer function $\mathbf{C}(\omega)$, which leads to an unstable adaptive algorithm. The second is the spectral deformation caused by $|\mathbf{C}(\omega)|$. The effect of the latter may be explained by the extreme case of each of $\{\mathbf{C}_{ml}(\omega_0) \ : \ l = 1, 2, \cdots, L\}$ has a zero response at the same frequency ω_0. In this case, $e_m(n)$ contains no information about the filters' outputs at ω_0, resulting in an unobservable system at that frequency.

A stable adaptive algorithm with $\mathbf{C}(\omega) \neq \mathbf{I}$ may only be obtained if the above mentioned filtering effects are compensated. The MEFX algorithm discussed in Section 3.3.1 solves this problem by filtering $\underline{\mathbf{x}}(n)$ through estimates of $\mathbf{C}(\omega)$ prior using them as inputs to the adaptation process as shown in Fig. 3.2. This effectively delays the inputs to be correctly aligned in time with the error signals. Since the dimensions of matrixes may not match and matrixes do not commute, the MEFX effectively filters every input signal $\underline{\mathbf{x}}_k(n)$ with every electro-acoustic response $\underline{\mathbf{c}}_{ml}$ to construct the KLM elements of the matrix $\mathbf{x}_f(n)$. This requires in total KLM convolution operations only for constructing $\mathbf{x}_f(n)$. Assuming time domain implementation, the total number of multiplications required per iteration for calculating $\mathbf{x}_f(n)$ is $KLMN_c$. Updating the weights requires $KLN_w(M+1)$ multiplications, and calculating the filters' outputs requires KLN_w multiplications. The total number of multiplications required to implement the MEFX algorithm each iteration is, therefore,

$$\Psi_{MEFX} = K\,L\,[(N_w + N_c)\,M + 2\,N_w], \tag{3.62}$$

which increases rapidly with increasing system dimensions. A much more efficient algorithm may be obtained by advancing the error signals $\underline{\mathbf{e}}(n)$ in time to be aligned with $\mathbf{x}(n)$. This leads to the Adjoint Least Mean Square (ALMS) algorithm [141], which has its update equation given by

$$\underline{\mathbf{w}}(n+1) = \underline{\mathbf{w}}(n) + 2\,\mu\,\mathbf{x}^T(n - N_c + 1)\ \underline{\mathbf{e}}_f(n - N_c + 1), \tag{3.63}$$

where $\underline{\mathbf{e}}_f = \begin{bmatrix} \underline{\mathbf{e}}_{f_1} & \underline{\mathbf{e}}_{f_2} & \cdots & \underline{\mathbf{e}}_{f_L} \end{bmatrix}^T$ is the result of filtering the error signals through the *adjoint* (time mirror) of the transfer functions comprising $\mathbf{C}(\omega)$. This non-causal operation can only be performed in real-time after a delay of $N_c - 1$ samples, the reason for using delayed input and

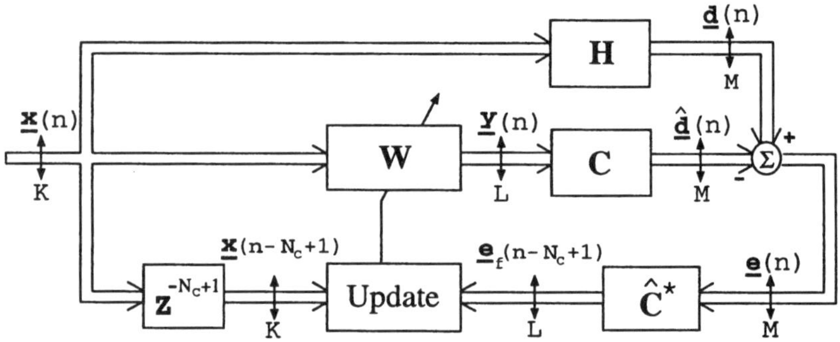

Figure 3.4: Block diagram of the adjoint least mean square algorithm.

filtered error signals in (3.63). The time mirror operation is equivalent to calculating the complex conjugate of the frequency response as shown in Fig. 3.4. Expressing the adjoint of $\underline{\mathbf{c}}_{ml}$ as

$$\check{\underline{\mathbf{c}}}_{ml} = \left[\; \underline{\mathbf{c}}_{ml,N_c-1} \;\; \underline{\mathbf{c}}_{ml,N_c-2} \;\; \cdots \;\; \underline{\mathbf{c}}_{ml,1} \;\; \underline{\mathbf{c}}_{ml,0} \;\right]^T, \tag{3.64}$$

the filtered error signal e_{f_l} may be written as by

$$\underline{\mathbf{e}}_{f_l}(n - N_c + 1) = \check{\underline{\mathbf{c}}}_{1l}\,\underline{\mathbf{e}}_1^T(n) + \check{\underline{\mathbf{c}}}_{2l}\,\underline{\mathbf{e}}_2^T(n) + \cdots + \check{\underline{\mathbf{c}}}_{Ml}\,\underline{\mathbf{e}}_M^T(n). \tag{3.65}$$

The ALMS algorithm reduces the number of multiplications required for updating the adaptive weights to $2KLN_w$. Adding LMN_c multiplications for calculating $\underline{\mathbf{e}}_f(n)$, and KLN_w multiplications for calculating the filters' outputs sums to

$$\Psi_{ALMS} = K\,L\,[\frac{N_c}{K}\,M + 3\,N_w]. \tag{3.66}$$

Comparing (3.66) with (3.62) shows that for a $[K \times L \times M = 1 \times 1 \times 1]$ system, the ALMS and MEFX have the same complexity of $(3\,N_w + N_c)$ multiplications. For multichannel systems, however, the ALMS algorithm is much more efficient than the MEFX as shown in Fig. 3.5, which gives the computational saving per iteration $(\Psi_{MEFX} - \Psi_{ALMS})$ against the adaptive filter order N_w for different system configurations.

A detailed implementation of the ALMS algorithm may be seen from the update equation of an individual filter $\underline{\mathbf{w}}_{lk}(n)$, which is given by

$$\underline{\mathbf{w}}_{lk}(n+1) = \underline{\mathbf{w}}_{lk}(n) + 2\,\mu\,e_{f_l}(n - N_c + 1)\;\underline{\mathbf{x}}_k(n - N_c + 1). \tag{3.67}$$

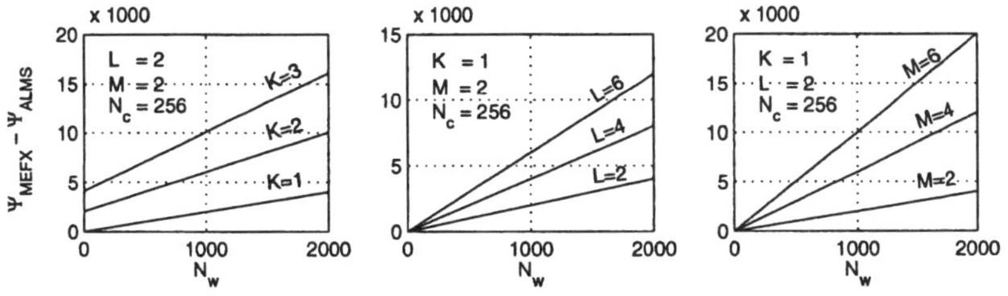

Figure 3.5: Computational saving per iteration ($\Psi_{MEFX} - \Psi_{ALMS}$) against the adaptive filter order N_w for different system configurations.

This update is shown in Fig. 3.6 for a $[K \times L \times M = 2 \times 2 \times 2]$ system in an active noise control setting similar to that in Fig. 3.3. A comparison of Fig. 3.6 and Fig. 3.3 reveals the structure simplicity offered by the ALMS over the MEFX in addition to the computational saving mentioned above. Two sources of computational saving may be recognised in this example. The first is the reduction of the number of convolutions required to calculate the filtered signals (4 in case of ALMS against 8 in MEFX). The second is the simplified ALMS update that always uses a single error signal and a single input signal compared to 2 error and 2 input signals in the MEFX case.

3.4.2 Frequency Domain Implementations

Implementation of the MEFX (Fig. 3.2) or ALMS (Fig. 3.4) requires performing three main tasks:

1. Calculating the adaptive filters' outputs $\underline{\mathbf{y}}(n)$.
2. Calculating the filtered inputs $\mathbf{x}_f(n)$ (or errors $\underline{\mathbf{e}}_f(n)$).
3. Updating the coefficients of the adaptive filters $\underline{\mathbf{W}}$.

The first two tasks are convolution operations between $\underline{\mathbf{x}}(n)$ and FIR filters comprising $\underline{\mathbf{W}}(\omega)$ and $\mathbf{C}(\omega)$, respectively. The third task implies calculating the instantaneous gradient that is essentially a crosscorrelation between the input and error signals. Since all FIR filters are of high order, implementing the convolution and correlation in the frequency domain results in considerable computational saving [114, 134]. Since the

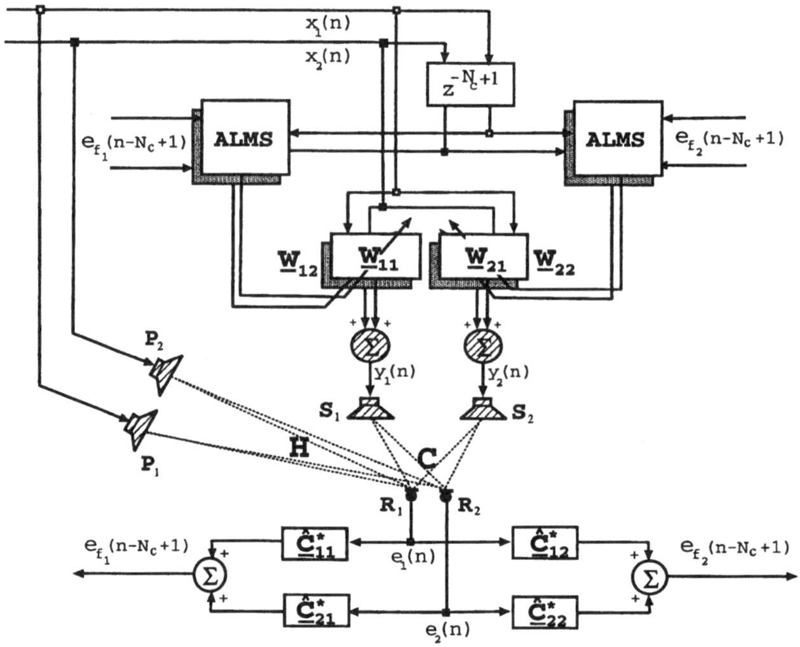

Figure 3.6: Block diagram of the ALMS algorithm for a $[K \times L \times M = 2 \times 2 \times 2]$ system in an active noise control setting.

input signals are infinitely long, real-time implementation of the convolution or correlation in the frequency domain must be performed on successive overlapping blocks of data. Two known, and frequently used, methods to construct and process such overlapping blocks of data are the overlap-add and overlap-save methods [114]. Applying any of those block processing techniques to the LMS adaptive process leads to the Block Frequency Domain Adaptive Filter (BFDAF) [41, 55, 98, 106, 130].

Besides the speed gained by carrying out the convolution and correlation operations in the frequency domain, BFDAF offers the possibility of approximate decorrelation of the adaptive process from the input signals' statistics. This is done by normalising each frequency bin by the input power in that bin [130]. Unlike the systems discussed previously which update the coefficients of the adaptive filters every sample, BFDAF performs such update every $L \geq 1$ samples, which leads to lower complexity, but at the same time slower convergence rate. Due to block processing, reduction in the step size bound and processing delay of L samples are

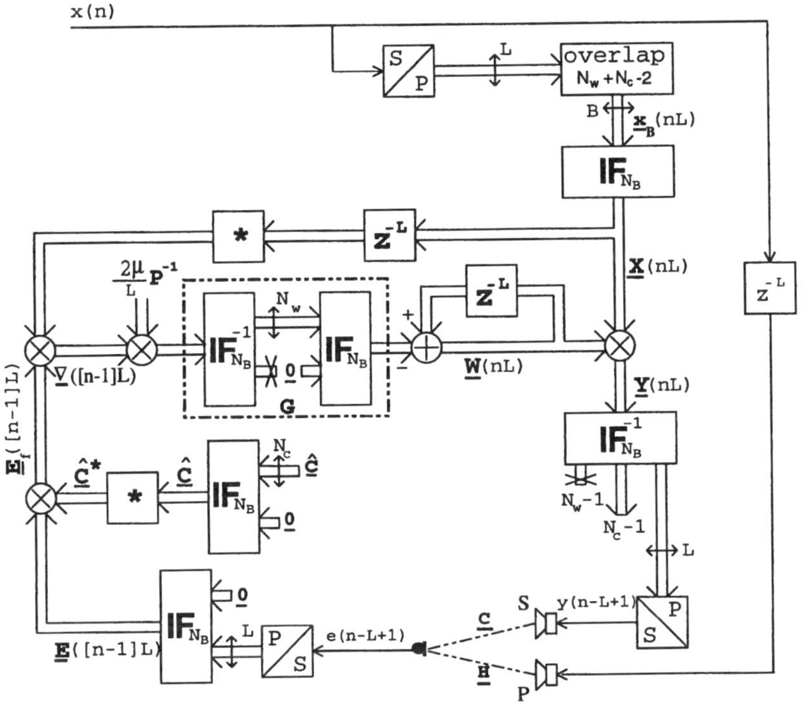

Figure 3.7: Block frequency domain implementation of the ALMS algorithm for a single channel $[K \times L \times M = 1 \times 1 \times 1]$ system.

also introduced. In this section, implementations of 3D sound systems using BFDAF are considered. Since the adjoint LMS algorithm is more efficient for multichannel systems as shown in Section 3.4.1, only BFDAF implementation of that algorithm is considered below.

For clarity, the BFDAF implementation of the ALMS algorithm is illustrated for a single channel active noise control system, since the extension to multichannel systems is straightforward as given by (3.63). The block diagram of the ALMS algorithm implemented using overlap-save BFDAF for $[K \times L \times M = 1 \times 1 \times 1]$ ANC system is shown in Fig. 3.7. In this block diagram, vectors are represented by double parallel lines while scalars are represented by single lines. Since we are dealing with a single channel case, vectors are used here to represent frequency samples of the signals rather than multiple signals at the same frequency.

The overlap-save method requires sectioning the stream of input time samples $x(n)$ into overlapping blocks, each of length N_B. Each block contains L new samples and $N_w + N_c - 2$ samples from the previous block. The sectioning and overlapping operations are represented in Fig. 3.7 by the *serial-to-parallel* and *overlap* blocks. Each L samples, a complete block of time samples $\underline{\mathbf{x}}_B(nL)$ is constructed and processing of this block of data may be commenced. For correct real-time operation, the processing of each block must be completed before the next block is constructed, i.e. in maximum L samples. To perform convolution and correlation in the frequency domain, each block $\underline{\mathbf{x}}_B(nL)$ is transformed to the frequency domain using a length $N_B = N_w + N_c + L - 2$ Fast Fourier Transform (FFT) algorithm, $\underline{\mathbf{X}}(nL) = \mathbb{F}_{N_B}\ \underline{\mathbf{x}}_B(nL)$. The length of the block N_B is chosen so that the circular correlation performed by element-wise multiplication in the frequency domain results in N_w correct time domain correlation coefficients as will be shown shortly. The processing of each block can be divided into three main tasks as mentioned above:

1. **Calculation of the filter's output** $y(n-L+1)$**:** The convolution between the transformed input block $\underline{\mathbf{X}}(nL)$ and the frequency domain adaptive filter weights $\underline{\mathbf{W}}(nL)$ is carried out by element-wise multiplication. Defining the matrix $\mathbf{X}(nL) = \text{diag}\{\underline{\mathbf{X}}(nL)\}$, this convolution may be represented by

$$\underline{\mathbf{Y}}(nL) = \mathbf{X}(nL)\,\underline{\mathbf{W}}(nL). \tag{3.68}$$

The vector $\underline{\mathbf{Y}}(nL)$ is then transformed back to the time domain using an inverse Fast Fourier transform (IFFT) algorithm of length N_B. The last L samples of this time domain vector represent the required linear convolution. During processing the next block, these L samples are sent, one sample each clock cycle, to drive loudspeaker S. This is represented in Fig. 3.7 by the *parallel-to-serial* block. The first $N_w - 1$ samples represent cyclic convolution results and must be discarded. An extra $N_c - 1$ samples represent correct linear convolution results but are not used since only L samples are needed to be sent out every block. This extra $N_c - 1$ samples are due to the choice of a large block length N_B to accommodate successive convolution and correlation as described below.

2. **Calculation of the filtered error** $\underline{\mathbf{E}}_f([n-1]L)$**:** The microphone signal $e(n - L + 1)$ is buffered into a length L vector. This vec-

tor is augmented with $N_B - L$ leading zeros and transformed to the frequency domain using an FFT of length N_B, resulting in the block error vector $\underline{\mathbf{E}}([n-1]L)$. A previously measured impulse response $\hat{\underline{\mathbf{c}}}$ of length N_c between loudspeaker S and the microphone is padded with $N_B - N_c$ trailing zeros and transformed to the frequency domain using an FFT of length N_B, producing the vector $\hat{\underline{\mathbf{C}}}$. The adjoint (time mirror) operation required for the ALMS is performed by calculating the complex conjugate of $\hat{\underline{\mathbf{C}}}$. The filtered error vector $\underline{\mathbf{E}}_f([n-1]L)$ is then calculated by element-wise multiplication of $\hat{\underline{\mathbf{C}}}^*$ and $\underline{\mathbf{E}}([n-1]L)$. The length of the result of this convolution is $L + N_c - 1$, and since N_B is chosen larger than this length, all samples of $\underline{\mathbf{E}}_f([n-1]L)$ represent correct linear convolution results. Defining the matrix $\hat{\mathbf{C}} = \text{diag}\{\hat{\underline{\mathbf{C}}}\}$, this convolution may be expressed mathematically as

$$\underline{\mathbf{E}}_f([n-1]L) = \hat{\mathbf{C}}^* \, \underline{\mathbf{E}}([n-1]L). \tag{3.69}$$

Since the microphone signal $e(n-L+1)$ is due to output samples $y(n-L+1)$ from the previous block, the error signal is already delayed by L samples. Provided that $L \geq N_c$, this delay will suffice to implement the delay required by the ALMS as mentioned in Section 3.4.1. To further ensure that the optimum filter solution is causal, the delay in the path through loudspeaker P must be longer than that through loudspeaker S. Since the algorithm introduces a delay of L samples, this delay must be compensated. Assuming that the acoustic transfer function $\underline{\mathbf{H}}$ possesses longer delay than $\underline{\mathbf{C}}$, this compensation may be done by delaying the signal to loudspeaker P by L samples. This is represented in Fig. 3.7 by the box z^{-L}. An elegant alternative is to move loudspeaker P an equivalent distance away from the microphone to save DSP memory.

3. **Updating the filter weights:** Transforming the update equation of the ALMS (3.63) to the frequency domain results in the following

block frequency domain update equation[7]

$$\underline{\mathbf{W}}(nL) = \underline{\mathbf{W}}([n-1]L) - \frac{2\mu}{L}\mathbf{G}\mathbf{P}^{-1}\,\mathbf{X}^{*}([n-1]L)\;\underline{\mathbf{E}}_f([n-1]L), \tag{3.70}$$

where it is assumed in this equation that $L \geq N_c$, the delay required by the ALMS algorithm. The diagonal matrix **P** represents an estimate of the input signal power in each frequency bin and performs the decorrelation mentioned above. The inverse of this power matrix is scaled by the step size $2\mu/L$ and the result is used as the new step size. This frequency-dependent step size vector is used to scale the estimated gradient vector. The instantaneous block mean square gradient estimate vector given by the cross-correlation $\underline{\nabla}([n-1]L) = \mathbf{X}^{*}([n-1]L)\;\underline{\mathbf{E}}_f([n-1]L)$ is obtained by first delaying $\underline{\mathbf{X}}(nL)$, calculating its complex conjugate, and element-wise multiplying the result by the block filtered error as shown in Fig. 3.7. Since $\underline{\mathbf{X}}([n-1]L)$ is not padded with zeros, this element-wise multiplication corresponds in the time domain to N_w linear correlation coefficients, while the last $N_B - N_w$ samples are the cyclic correlation coefficients and must be discarded. However, the correct correlation coefficients must be extracted in the time domain, which is done using the $N_B \times N_B$ constraining window **G**. This window transforms the estimated gradient to the time domain, replaces the last $N_B - N_w$ samples by zeros, and transforms the result back to the frequency domain. Finally, the weighted and constrained gradient is used to update the weight vector $\underline{\mathbf{W}}([n-1]L)$ according to (3.70).

The constraining window **G** that costs two FFTs may be dropped if extra computational saving is needed. In this case, the algorithm is referred to as the unconstrained BFDAF [98, 130]. This leads to using cyclic correlation coefficients in the gradient vector, which in turn leads to a slower convergence rate and a reduction in the stable range of the step size [98]. Although the unconstrained filter corresponds to N_B coefficients in the time domain, the 'wrap-around error' prevents it from performing as a filter of more coefficients than N_w.

---

[7]Since signals are summed up at the microphones in Fig. 3.7 rather than subtracted, the + sign in (3.63) has been changed to a − sign in (3.70).

Finally, it is worth mentioning that the frequency domain implementation of a single channel system using the adjoint LMS may lead to more computational saving over that using the filtered-x algorithm if N_w is much larger than N_c. This second source of computational saving stems from the fact that the filtered-x algorithm calculates $\mathbf{x}_f$ by filtering the infinitely long input signal through the finite impulse response $\underline{\mathbf{C}}$. Implementing this convolution in the frequency domain requires the use of an overlapping method, and possibly two extra FFTs to separate the correct convolution samples in the time domain. On the other hand, the frequency domain adjoint LMS algorithm calculates $\underline{\mathbf{E}}_f$ by filtering (the already in block form) error signal through $\underline{\mathbf{C}}^*$ and no overlapping is needed. BFDAF implementations of the single channel adjoint LMS and several implementations of the single channel filtered-x have already been presented in previous works [3, 83, 100, 131, 140]. A complexity comparison between these implementations shows that for a single channel BFDAF, the filtered-x is more efficient than the adjoint LMS when $N_w < 2\,N_c$, while the adjoint LMS is much more efficient when $N_w > 2\,N_c$ [3, 131].

Chapter 4

Robustness of 3D Sound Systems

The solution for the control filters in the generalised audio reproduction model discussed in Section 3.1 consists of two elements. The first is the cross-talk canceller, which is the inverse of the matrix of electro-acoustic transfer functions between the reproduction loudspeakers and the listeners' ears. The second is the matrix of electro-acoustic transfer functions representing the desired response. Since both elements are dependent on space coordinates, the solution for the control filters is inherently dependent on space coordinates, independent of the method used to calculate the filters. A set of filters designed to give optimum results at a specific listening situation is, in general, in error when a listener moves. In this chapter, the nature of those errors is discussed and methods to improve the system robustness against listeners' movements are introduced. The problem is divided into two subproblems: the first deals with enlarging the zones of equalisation created around the control microphones and the second tracks the listeners when they move beyond those equalisation zones.

In Section 4.1, the physical nature of the errors introduced in the control filters when a listener moves is discussed. These errors are not only position dependent, but also frequency dependent. Small movements make the control filters slightly in error at low frequencies while the errors increase with increasing frequency. This motivates using control filters having non-uniform frequency resolution to cope with this frequency dependency.

The equalisation zones may be enlarged by designing the control filters to be valid not only at the eardrums of the listeners but also at points in space in the vicinity of the listeners' ears. This is achieved by constraining the derivative of the sound field with respect to the space coordinates in the proximity of the listeners' eardrums to be zero. The implementation of derivative constrained filters requires measuring several transfer functions in the vicinity of the listeners' ears and designing the filters using a weighted average of those. Methods for designing the filters to meet this constraint are discussed in Section 4.2. These methods achieve their goal by forcing the filters to assume solutions that increasingly deviate from their exact solutions with increasing frequency, which is a second motivation for using control filters having non-uniform frequency resolution. By decreasing the frequency resolution of the control filters as the frequency increases, the filters ignore more details with increasing frequency. This makes the filters valid in a wider area in space without the need to measure several transfer functions at adjacent points. Multiresolution control filters are discussed in Section 4.3 and in more details in Chapter 5.

The zones of equalisation can further be improved by using more reproduction loudspeakers to increase system controllability. This is achieved only if the reproduction loudspeakers are correctly positioned in the listening space. Improving system controllability by employing more reproduction loudspeakers and the choice of the positions of those loudspeakers are discussed in Section 4.4 and Section 4.5, respectively.

Large errors are introduced in the control filters when the listeners move beyond the equalisation zones. In case of such large movements, the filters must be recalculated using the transfer functions corresponding to the listeners' new positions. In adaptive 3D sound systems, this may be achieved by readapting the control filters to their new optimal solutions. This in turn requires measuring all electro-acoustic transfer functions between the reproduction loudspeakers and the listeners' ears that are needed for the filtered-x and adjoint LMS adaptive algorithms discussed in Chapter 3. Large movements and head tracking are discussed in Section 4.6, while on-line estimation of the electro-acoustic transfer functions and on-line adaptation of the control filters are discussed in

Section 4.7.

Essential for adaptive 3D sound systems is that the adaptive filters reach their new optimum solutions after a listener movement fast enough so that the listener does not loose the sound image. Since the above mentioned adaptive algorithms are known for their slow convergence speed, this requires improving the convergence properties of these algorithms, which is addressed in Section 4.8. Finally, Section 4.9 summarises the techniques introduced in this chapter.

4.1 The Robustness Problem

In the general sound reproduction model shown in Fig. 3.1, the $[L \times K]$ matrix of control filters $\mathbf{W}(\omega)$ that minimises the error vector $\underline{\mathbf{E}}(\omega)$ for a given desired response $\underline{\mathbf{D}}(\omega)$ is obtained by solving the system of control equations

$$\mathbf{C}(\omega)\,\mathbf{W}(\omega) = \mathbf{H}(\omega), \tag{4.1}$$

where $\mathbf{C}(\omega)$ is the $[M \times L]$ matrix of transfer functions between the loudspeakers $\{S_1, S_2, \cdots S_L\}$ and the microphones $\{R_1, R_2, \cdots R_M\}$ at ω. The $[M \times K]$ matrix $\mathbf{H}(\omega)$ is used to generate the desired response vector

$$\underline{\mathbf{D}}(\omega) = \mathbf{H}(\omega)\,\underline{\mathbf{X}}(\omega). \tag{4.2}$$

Regardless of the method used to calculate the filters $\mathbf{W}(\omega)$, it is clear from (4.1) that the solution is dependent only on the transfer functions $\{\underline{\mathbf{C}}_{ml}(\omega) : m = 1, 2, \cdots, M,\ l = 1, 2, \cdots, L\}$ and $\{\underline{\mathbf{H}}_{mk}(\omega) : m = 1, 2, \cdots, M,\ k = 1, 2, \cdots, K\}$. Each of those transfer functions is a cascade combination of several components. These components include:

- The impulse responses of the reconstruction filter and the Digital-to-Analogue Converter (DAC) used to transform $\underline{\mathbf{y}}(n)$ and $\underline{\mathbf{x}}(n)$ to continuous-time form.
- The impulse responses of the power amplifiers used to drive the loudspeakers.
- The impulse responses of the loudspeakers $\{S_1, S_2, \cdots, S_L\}$ for $\mathbf{C}(\omega)$ and the primary loudspeakers $\{P_1, P_2, \cdots, P_K\}$ for $\mathbf{H}(\omega)$. For some applications, such as cross-talk cancellation, the primary

loudspeakers do not physically exist and their responses may be considered as delta functions.

- The room impulse responses between each of the above mentioned loudspeakers and each microphone.
- The impulse responses of the microphones and their corresponding amplifiers.
- The Head-Related Transfer Functions (HRTFs) from each loudspeaker to each of the listeners' ears.
- The responses of the anti-aliasing filters and the Analogue-to-Digital Converters (ADC) that transform the microphones' signals to digital form.

From all these components, the head-related transfer functions and the impulse responses between points in the room are functions in space coordinates as mentioned in Chapter 2. Consequently, the optimal solution for $\mathbf{W}(\omega)$ is dependent on the position of the microphones relative to the loudspeakers. This position dependency is further frequency dependent due to the slow speed at which sound waves propagate in the air. To illustrate this latter statement, consider a 2 cm displacement in one of the microphones away from its original position. This displacement corresponds to a complete wavelength of a 17 kHz sound wave while it is just 0.01 wavelength of a 170 Hz wave. A matrix of filters $\mathbf{W}(\omega)$ designed for optimal solution at the original microphones' positions has higher amplitude errors at higher frequencies than at lower frequencies at the displaced microphone position.

The frequency dependent spatial dependency mentioned above is illustrated in Fig. 4.1. This result was obtained in a single channel active noise control experiment in an anechoic chamber [52, 53, 65]. In this experiment, white noise filtered by a band-pass filter having lower and higher cut-off frequencies of 0.1 and 4 kHz, respectively, is emitted from a primary loudspeaker. An adaptive filter is used to drive a secondary source to cancel the sound at a microphone placed 0.5 m from the secondary source. From Fig. 4.1, substantial decrease in the primary acoustic power is achieved at all frequencies at the microphone position. At points away from this control position, the reduction in primary acoustic power decreases. This decrease is frequency dependent; at the same

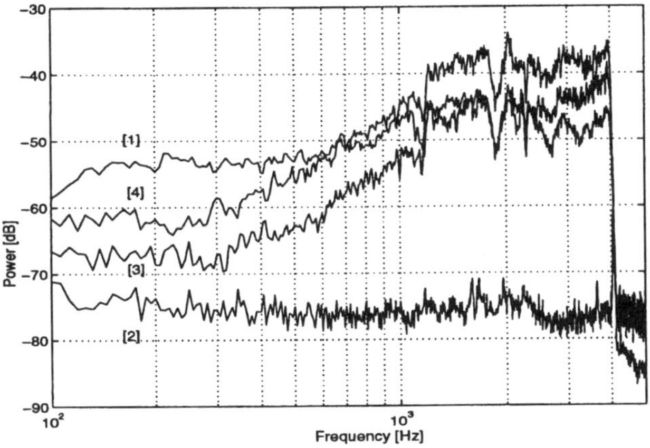

Figure 4.1: The acoustic power spectrum measured at different points from the control microphone in a single channel active noise control experiment. Curve [1] is the primary noise to be reduced, curve [2] is the resulting acoustic power after cancellation at the microphone position, [3] is that at 2 cm from the microphone, and [4] is at 5 cm from the microphone.

point in space, a rapid decrease in the sound reduction is noticed as the frequency increases. At 2 cm from the control microphone, only about 5 dB reduction of primary acoustic power is achieved at 1 kHz, which is not satisfactory for any audio reproduction application. At the end of the audible frequency range (at 20 kHz), one expects that even a few millimetres displacement in one of the microphones or loudspeakers may cause enough errors that render the performance of the system not acceptable.

In general, loudspeakers and microphones also have directional characteristics, which further complicates the matter. In the present work, the loudspeakers and microphones are considered omnidirectional to simplify the discussion. An omnidirectional microphone produces the same output voltage for an incident sound wave regardless of the arrival angle of the sound. Similarly, an omnidirectional loudspeaker radiates equally strong sound waves in all directions.

4.2 Enlarging the Zones of Equalisation

The control task of the generalised model shown in Fig. 3.1 may be described by the system of control equations

$$\mathbf{C}(\omega, r, \alpha, \theta)\ \mathbf{W}(\omega) = \mathbf{H}(\omega, r, \alpha, \theta), \tag{4.3}$$

where $\mathbf{C}(\omega, r, \alpha, \theta)$ is the $[M \times L]$ matrix of electro-acoustic transfer functions between the M microphones and the L reproduction loudspeakers. $\mathbf{H}(\omega, r, \alpha, \theta)$ is the $[M \times K]$ matrix of electro-acoustic transfer functions between the M microphones and the K input signals. $\mathbf{W}(\omega)$ is the $[L \times K]$ matrix of control filters. Since (4.3) does not specify the sound field at points other than the control points (the positions of the microphones), direct solution of (4.3) results in control filters that accurately fulfil the control task only at the positions of the microphones. The sound field at points away from the control points is, therefore, uncontrollable by the filters. This leads to small equalisation zones around the microphones as mentioned above, which makes the system sensitive to changes in the microphones' positions. To enlarge the equalisation zones, the control filters must be designed to control the sound field not only at, but also in the vicinity of, the microphones. Extra constraints that specify the sound field in the vicinity of the microphones must, therefore, be added to the system of control equations (4.3). Several of such constraints are discussed in Sections 4.2.1 through 4.2.3. The performance of the discussed methods is examined by computer experiments in Section 4.2.4

4.2.1 Spatial Derivative Constraints

One approach to enlarge the equalisation zones is to design the control filters $\mathbf{W}(\omega)$ such that the actual sound field $\hat{\underline{\mathbf{D}}}(\omega)$ is constant and equal to the desired sound field $\underline{\mathbf{D}}(\omega)$ not only at the microphones but also in their proximity. This is equivalent to constraining the first derivative of $\mathbf{C}(\omega, r, \alpha, \theta)\,\mathbf{W}(\omega)$ with respect to the space coordinates to zero. Further expansion of the control action in space may be obtained by constraining the first $J > 1$ spatial derivatives to zero [10]. Derivative constraints with respect to the vertical-polar coordinates allow designing the filters $\mathbf{W}(\omega)$ while taking into account radial movements in the r direction, and head rotations in α and θ directions. Alternatively, the Cartesian coordinates system may be used, where derivative constraints minimise errors in the filters due to translations in the x, y, and z directions.

Spatial derivative constraints are applied by adding new control equations to the set in (4.3), each of which constrains the derivative of the sound field near the corresponding microphone to be zero. Singling out the control equation for the mth microphone positioned at coordinates $(r_m, \alpha_m, \theta_m)$ due to the kth input signal in (4.3), and dropping the dependency on ω and space coordinates for improved readability, we obtain

$$C_{m1}W_{1k} + C_{m2}W_{2k} + \cdots + C_{mL}W_{Lk} = H_{mk}. \tag{4.4}$$

The corresponding first-order spatial derivative constrained control equations with respect to r_m, α_m, and θ_m are

$$\sum_{l=1}^{L} \frac{\partial C_{ml}(\omega, r_m, \alpha_m, \theta_m)}{\partial r_m} W_{lk}(\omega) = 0, \tag{4.5}$$

$$\sum_{l=1}^{L} \frac{\partial C_{ml}(\omega, r_m, \alpha_m, \theta_m)}{\partial \alpha_m} W_{lk}(\omega) = 0, \tag{4.6}$$

$$\sum_{l=1}^{L} \frac{\partial C_{ml}(\omega, r_m, \alpha_m, \theta_m)}{\partial \theta_m} W_{lk}(\omega) = 0. \tag{4.7}$$

The first-order spatial derivative constrained control equations for all microphones can, therefore, be written as

$$\overline{\mathbf{C}}(\omega, r, \alpha, \theta)\ \mathbf{W}(\omega) = \overline{\mathbf{H}}(\omega, r, \alpha, \theta), \tag{4.8}$$

where $\overline{\mathbf{C}}(\omega, r, \alpha, \theta)$ and $\overline{\mathbf{H}}(\omega, r, \alpha, \theta)$ are given by

$$\overline{\mathbf{C}} = \begin{bmatrix} C_{11} & \cdots & C_{1L} \\ \partial C_{11}/\partial r_1 & \cdots & \partial C_{1L}/\partial r_1 \\ \partial C_{11}/\partial \alpha_1 & \cdots & \partial C_{1L}/\partial \alpha_1 \\ \partial C_{11}/\partial \theta_1 & \cdots & \partial C_{1L}/\partial \theta_1 \\ \vdots & \ddots & \vdots \\ C_{M1} & \cdots & C_{ML} \\ \partial C_{M1}/\partial r_M & \cdots & \partial C_{ML}/\partial r_M \\ \partial C_{M1}/\partial \alpha_M & \cdots & \partial C_{ML}/\partial \alpha_M \\ \partial C_{M1}/\partial \theta_M & \cdots & \partial C_{ML}/\partial \theta_M \end{bmatrix}, \quad \overline{\mathbf{H}} = \begin{bmatrix} H_{11} & \cdots & H_{1K} \\ 0 & \cdots & 0 \\ 0 & \cdots & 0 \\ 0 & \cdots & 0 \\ \vdots & \ddots & \vdots \\ H_{M1} & \cdots & H_{MK} \\ 0 & \cdots & 0 \\ 0 & \cdots & 0 \\ 0 & \cdots & 0 \end{bmatrix}. \tag{4.9}$$

Higher order spatial derivative constraints are applied by further adding the higher order derivative control equations and equating each to zero [10]. Implementation of the spatial derivative constrained filters requires analytic expressions for the electro-acoustic transfer functions $\mathbf{C}(\omega, r, \alpha, \theta)$ for calculating the spatial derivatives. Such analytic expressions are readily calculated in free field listening conditions (with the listener absent), where the transfer functions are dependent only on the distance between the microphones and the reproduction loudspeakers. Analytical expressions may also be obtained for the direct sound and the first few reflections in cases of moderate reverberant conditions. As the reverberation level increases, it becomes more difficult to derive the required analytical expressions and approximations to the spatial derivatives must be employed. The performance of such approximate spatial derivative constrained filters is examined by computer simulations in Section 4.2.4.3.

4.2.2 Spatial Difference Constraints

As mentioned above, exact spatial derivative constraints may be applied in free field conditions, where analytical expressions for the electro-acoustic transfer functions can be readily calculated. In practical reverberant environments, however, only approximate calculations of the derivatives are possible. One such approximation is controlling the sound field at discrete points in the proximity of the original control microphones [4, 5, 10]. In this approach, extra control microphones are placed at small displacements from the original control points. The control filters are then designed to control the sound field at the displaced microphones to be the same as the desired sound field at the original control points.

Consider adding three microphones at $(r_m + \Delta r_m, \alpha_m, \theta_m)$, $(r_m, \alpha_m + \Delta\alpha_m, \theta_m)$, and $(r_m, \alpha_m, \theta_m + \Delta\theta_m)$ in the proximity of the original m^{th} control point located at $(r_m, \alpha_m, \theta_m)$, where Δr_m, $\Delta\alpha_m$, and $\Delta\theta_m$ are small displacements from the original position of the mth microphone in the corresponding space coordinate directions. The control equations for those four microphones due to the kth source can be written as

$$\sum_{l=1}^{L} C_{ml}(\omega, r_m, \alpha_m, \theta_m) W_{lk}(\omega) = H_{mk}(\omega, r_m, \alpha_m, \theta_m), \qquad (4.10)$$

$$\sum_{l=1}^{L} C_{ml}(\omega, r_m + \Delta r_m, \alpha_m, \theta_m) W_{lk}(\omega) = H_{mk}(\omega, r_m, \alpha_m, \theta_m), \quad (4.11)$$

$$\sum_{l=1}^{L} C_{ml}(\omega, r_m, \alpha_m + \Delta\alpha_m, \theta_m) W_{lk}(\omega) = H_{mk}(\omega, r_m, \alpha_m, \theta_m), \quad (4.12)$$

$$\sum_{l=1}^{L} C_{ml}(\omega, r_m, \alpha_m, \theta_m + \Delta\theta_m) W_{lk}(\omega) = H_{mk}(\omega, r_m, \alpha_m, \theta_m). \quad (4.13)$$

Subtracting each of (4.11), (4.12), and (4.13) from (4.10) and dividing by the corresponding displacement results in

$$\sum_{l=1}^{L} \frac{C_{ml}(\omega, r_m + \Delta r_m, \alpha_m, \theta_m) - C_{ml}(\omega, r_m, \alpha_m, \theta_m)}{\Delta r_m} W_{lk}(\omega) = 0, \quad (4.14)$$

$$\sum_{l=1}^{L} \frac{C_{ml}(\omega, r_m, \alpha_m + \Delta\alpha_m, \theta_m) - C_{ml}(\omega, r_m, \alpha_m, \theta_m)}{\Delta\alpha_m} W_{lk}(\omega) = 0, \quad (4.15)$$

$$\sum_{l=1}^{L} \frac{C_{ml}(\omega, r_m, \alpha_m, \theta_m + \Delta\theta_m) - C_{ml}(\omega, r_m, \alpha_m, \theta_m)}{\Delta\theta_m} W_{lk}(\omega) = 0. \quad (4.16)$$

Comparing (4.5), (4.6), and (4.7) with (4.14), (4.15), and (4.16), respectively, shows that the latter three control equations approximate the former three by replacing the spatial derivatives by their corresponding spatial differences. The first-order spatial difference constrained control equations for the whole system becomes

$$\widetilde{\mathbf{C}}(\omega, r, \alpha, \theta)\ \mathbf{W}(\omega) = \widetilde{\mathbf{H}}(\omega, r, \alpha, \theta), \quad (4.17)$$

where $\widetilde{\mathbf{C}}(\omega, r, \alpha, \theta)$ and $\widetilde{\mathbf{H}}(\omega, r, \alpha, \theta)$ are given by

$$\widetilde{\mathbf{C}} = \begin{bmatrix} C_{11}(\omega, r_1, \alpha_1, \theta_1) & \cdots & C_{1L}(\omega, r_1, \alpha_1, \theta_1) \\ C_{11}(\omega, r_1 + \Delta r_1, \alpha_1, \theta_1) & \cdots & C_{1L}(\omega, r_1 + \Delta r_1, \alpha_1, \theta_1) \\ C_{11}(\omega, r_1, \alpha_1 + \Delta\alpha_1, \theta_1) & \cdots & C_{1L}(\omega, r_1, \alpha_1 + \Delta\alpha_1, \theta_1) \\ C_{11}(\omega, r_1, \alpha_1, \theta_1 + \Delta\theta_1) & \cdots & C_{1L}(\omega, r_1, \alpha_1, \theta_1 + \Delta\theta_1) \\ \vdots & \ddots & \vdots \\ C_{M1}(\omega, r_M, \alpha_M, \theta_M) & \cdots & C_{ML}(\omega, r_M, \alpha_M, \theta_M) \\ C_{M1}(\omega, r_M + \Delta r_M, \alpha_M, \theta_M) & \cdots & C_{ML}(\omega, r_M + \Delta r_M, \alpha_M, \theta_M) \\ C_{M1}(\omega, r_M, \alpha_M + \Delta\alpha_M, \theta_M) & \cdots & C_{ML}(\omega, r_M, \alpha_M + \Delta\alpha_M, \theta_M) \\ C_{M1}(\omega, r_M, \alpha_M, \theta_M + \Delta\theta_M) & \cdots & C_{ML}(\omega, r_M, \alpha_M, \theta_M + \Delta\theta_M) \end{bmatrix} \tag{4.18}$$

$$\widetilde{\mathbf{H}} = \begin{bmatrix} H_{11}(\omega, r_1, \alpha_1, \theta_1) & \cdots & H_{1K}(\omega, r_1, \alpha_1, \theta_1) \\ H_{11}(\omega, r_1, \alpha_1, \theta_1) & \cdots & H_{1K}(\omega, r_1, \alpha_1, \theta_1) \\ H_{11}(\omega, r_1, \alpha_1, \theta_1) & \cdots & H_{1K}(\omega, r_1, \alpha_1, \theta_1) \\ H_{11}(\omega, r_1, \alpha_1, \theta_1) & \cdots & H_{1K}(\omega, r_1, \alpha_1, \theta_1) \\ \vdots & \ddots & \vdots \\ H_{M1}(\omega, r_M, \alpha_M, \theta_M) & \cdots & H_{MK}(\omega, r_M, \alpha_M, \theta_M) \\ H_{M1}(\omega, r_M, \alpha_M, \theta_M) & \cdots & H_{MK}(\omega, r_M, \alpha_M, \theta_M) \\ H_{M1}(\omega, r_M, \alpha_M, \theta_M) & \cdots & H_{MK}(\omega, r_M, \alpha_M, \theta_M) \\ H_{M1}(\omega, r_M, \alpha_M, \theta_M) & \cdots & H_{MK}(\omega, r_M, \alpha_M, \theta_M) \end{bmatrix}. \tag{4.19}$$

The matrixes $\widetilde{\mathbf{C}}$ and $\widetilde{\mathbf{H}}$ are composed of electro-acoustic transfer functions only, no derivative terms appear. This makes the system of control equations in (4.17) similar in nature to (4.3) except for the increase in the number of control microphones. Therefore, the control filters $\mathbf{W}(\omega)$ may be directly implemented using adaptive techniques as discussed in Chapter 3, while more control is achieved on the sound field. For sound reproduction applications, this means placing multiple microphones in the vicinity of each listener's ear.

In the previous discussion, only one extra microphone (displaced in each coordinate direction) has been used to approximate the derivative constraint in that coordinate direction. Better approximation of the first derivatives using 3-point or 5-point formulas [36] may be obtained by using more displaced microphones in each coordinate direction in the vicinity of each control point. More microphones can also be used to approximate higher order derivatives to further widen the zone of equalisation as mentioned in Section 4.2.1. Using two displaced microphones in the r direction at $(r_m - \Delta r_m, \alpha_m, \theta_m)$ and $(r_m + \Delta r_m, \alpha_m, \theta_m)$ in addition to the original microphone at $(r_m, \alpha_m, \theta_m)$, the second-order derivative of $C_{ml}(\omega, r_m, \alpha_m, \theta_m)$ with respect to r_m may be approximated by [36]

$$\frac{\partial^2 C_{ml}(\omega, r_m)}{\partial r_m^2} \approx \frac{C_{ml}(\omega, r_m + \Delta r_m) - 2C_{ml}(\omega, r_m) + C_{ml}(\omega, r_m - \Delta r_m)}{(\Delta r_m)^2}, \tag{4.20}$$

where the dependency on α_m and θ_m has been dropped in this equation for clarity. The second-order difference constrained control equation at the mth control point due to the kth sound source with respect to r_m may then be written as

$$\sum_{l=1}^{L} \frac{C_{ml}(\omega, r_m + \Delta r_m) - 2C_{ml}(\omega, r_m) + C_{ml}(\omega, r_m - \Delta r_m)}{(\Delta r_m)^2} W_{lk}(\omega) = 0. \tag{4.21}$$

Second-order difference constrained control equations with respect to θ^2, α^2, $r\alpha$, $r\theta$, and $\alpha\theta$ can be similarly obtained. As for the first-order case, second-order difference control equations are implemented by simply constraining the target transfer functions at all three microphones to be $H_{mk}(\omega, r_m, \alpha_m, \theta_m)$.

Since difference equations are correct discrete approximations to derivative equations only when the sampling theorem is honoured, the properties of the equalisation zones obtained using difference constraints depend on the spacing between adjacent microphones. Derivative constraints are equivalent to using an infinite number of control microphones around each of the control points, and smooth interpolation of the sound field at all frequencies is achieved. In the difference constrained case however, a single zone of equalisation is created around each microphone. The size of each zone decreases with increasing frequency similar to the behaviour shown in Fig. 4.1, and smooth interpolation at all frequencies is not guaranteed if the distances between adjacent microphones are large compared to the sound wavelength. At low frequencies, the size of the created zones is large enough to overlap and constitute one larger equalisation zone. As the frequency increases, the size of the individual zones decreases and the single zone starts to split leaving unequalised areas between the adjacent microphones. This behaviour of the zones of equalisation has also been observed in active noise control studies (where they are known as the *zones of quiet*) [57, 87, 109]. Experimental results in reverberant environments show that the zones of quiet around a microphone within which the error is reduced by 10 dB is a sphere of diameter 0.1 λ_{min}, where λ_{min} is the wavelength of the highest frequency of interest. This suggests that adjacent microphones must be at most 0.1 λ_{min} apart for a non-splitting zone of quiet of 10 dB error reduction.

In audio applications, the highest frequency of interest (20 kHz) corresponds to $\lambda_{min} \approx 17$ mm, and a separation of 1.7 mm between adjacent microphones is required for 10 dB equalisation. This reveals several deficiencies of the method of difference constraints for audio applications:

- The equalisation zones created around the microphones and, therefore, around the listener's ears, are too small at high frequencies. They allow the listener to move only a few millimetres.
- The separation distance between the microphones for a 10 dB equalisation zone at high frequencies is very small, even smaller than the physical diameter of most microphones.
- The 10 dB error reduction is not sufficient for most audio applications. Listening tests [51] show that at least 20 dB error reduction is required for audio only (opposed to audiovisual) applications. This means that an even smaller separation between the microphones is necessary[1].
- Every extra microphone adds L electro-acoustic transfer functions to the system. When the control filters are implemented as adaptive filters, this adds L identification filters and L convolutions to calculate the filtered input (or error) signals through the newly introduced transfer functions, which increases the system complexity considerably[2].

However, the difference constraint approach represents a powerful means of widening the equalisation zones at low and middle frequencies. This is most effective in applications such as active noise control of low frequency disturbances and speech applications. It can also be used to improve the

[1] Increasing the equalisation level by using more loudspeakers is addressed in Section 4.4.

[2] Reducing the complexity of difference constrained systems is addressed in Section 4.2.3 and 4.3.

system robustness at low frequencies while high frequencies are managed by other means. The performance of difference constrained filters is examined using computer simulations in Section 4.2.4.4.

4.2.3 Spatial Filters

As mentioned in Section 4.2.2, adding extra microphones in the proximity of the original control positions to apply the difference derivative constraints increases the system complexity considerably. Furthermore, increasing the number of microphones requires increasing the number of reproduction loudspeakers to obtain sufficient error reduction. One approach to contain this rapidly increasing complexity is to reduce the number of control equations. Rather than requiring the sound pressure at each extra control microphone to equal that desired at the adjacent original control point as in (4.17), a linear combination of such conditions may be employed. Such a linear combination reduces to spatially filtering all transfer functions around each control point. This reduces the number of control equations for each original control point to only two. The first equation is the original control equation and the second is a weighted average of all control equations corresponding to the displaced microphones in the vicinity of the original control point.

A linear combination for the conditions that all spatial differences up to the Jth-order to vanish also leads to the above mentioned spatial filter. This may be readily shown by considering the first and second difference control equations in the r direction given by (4.14) and (4.21), respectively. Multiplying (4.14) by a_1, (4.21) by a_2 and adding the results gives the weighted average spatial difference constrained control equation at the mth control point due to the kth sound source

$$\begin{aligned}\textstyle\sum_{l=1}^{L} [\, (a_1 + a_2)\, C_{ml}(\omega, r_m + \Delta r_m) - (a_1 + 2a_2)\, C_{ml}(\omega, r_m) \\ + a_2\, C_{ml}(\omega, r_m - \Delta r_m) \,]\, W_{lk}(\omega) = 0,\end{aligned} \tag{4.22}$$

where a_1 and a_2 are weighting coefficients for the first and second derivatives, respectively. Extending the above linear combination to include derivatives with respect to α_m and θ_m as well as r_m shows that the weighted average of the constraints simply reduces to the weighted average of the transfer functions themselves. This may be seen as a spatial filtering operation. For N microphones around the mth control point, this spatial filter is expressed as

$$\check{C}_{ml}(\omega, r_m, \alpha_m, \theta_m) = \sum_{n=1}^{N} b_n \, C_{nl}(\omega, r_m, \alpha_m, \theta_m). \tag{4.23}$$

In eq. (4.23), $C_{nl}(\omega, r_m, \alpha_m, \theta_m)$ is the electro-acoustic transfer function from the lth reproduction loudspeaker to the nth microphone near the mth original control point at ω, and b_n is the corresponding weighting coefficient. Using the notation for the spatial filtering result $\check{C}_{ml}$, the system of control equations may be written as

$$\begin{bmatrix} C_{11} & \cdots & C_{1L} \\ \check{C}_{11} & \cdots & \check{C}_{1L} \\ \vdots & \ddots & \vdots \\ C_{M1} & \cdots & C_{ML} \\ \check{C}_{M1} & \cdots & \check{C}_{ML} \end{bmatrix} \begin{bmatrix} W_{11} & \cdots & W_{1K} \\ \vdots & \ddots & \vdots \\ W_{L1} & \cdots & W_{LK} \end{bmatrix} = \begin{bmatrix} H_{11} & \cdots & H_{1K} \\ 0 & \cdots & 0 \\ \vdots & \ddots & \vdots \\ H_{M1} & \cdots & H_{MK} \\ 0 & \cdots & 0 \end{bmatrix}. \tag{4.24}$$

It is shown in Section 4.2.4.5 using computer simulations that although the system dimensions in (4.24) is much reduced compared to (4.17), suitable choice of the filter coefficients b_n results in equalisation zones that are comparable to those obtained using difference and approximate derivative constrained filters.

4.2.4 Computer Experiments

In this section, the performance of the above mentioned methods for enlarging the zones of equalisation is examined using computer simulations. Several implementation issues are also discussed.

4.2.4.1 Experimental Set-Up

Figure 4.2 shows the set-up used in the computer experiments discussed in this and next sections. Two loudspeakers S_1 and S_2 separated by a distance d_s are used to reproduce the audio signal $x(n)$. Two microphones R_1 and R_2 separated by a distance d_m are used to simulate the positions of a listener's ears when the listener is absent. The two loudspeakers

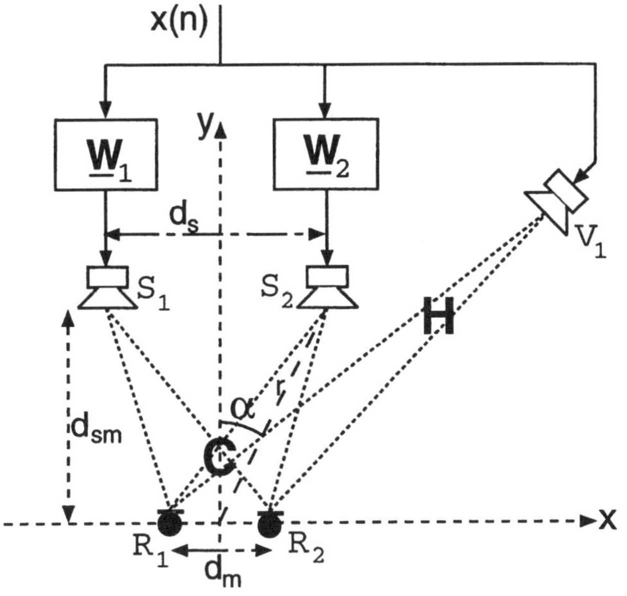

Figure 4.2: Experimental set-up.

are positioned on a straight line parallel to the line passing through the two microphones at a distance d_{sm}, so that each loudspeaker is at an azimuth angle $\alpha = \mathrm{atan}(d_s/2d_{sm})$. All loudspeakers and microphones are positioned in the horizontal plane $z = 1$ m.

The room simulation program *Room Impulse Response 2.1* [71] is used to calculate the four electro-acoustic transfer functions between the loudspeakers and the microphones. These transfer functions are labelled $\underline{\mathbf{C}}_{11}(\omega)$, $\underline{\mathbf{C}}_{12}(\omega)$, $\underline{\mathbf{C}}_{21}(\omega)$, and $\underline{\mathbf{C}}_{22}(\omega)$. The room simulation program is also used to calculate the desired transfer functions $\underline{\mathbf{H}}_{11}(\omega)$ and $\underline{\mathbf{H}}_{21}(\omega)$ from an arbitrarily chosen virtual source position V_1 to R_1 and R_2, respectively. Each transfer function is calculated as a FIR filter of 2048 taps at 44.1 kHz sampling frequency. The impulse responses are also convolved with a measured impulse response of a loudspeaker to account for the responses of the loudspeakers, microphones, DAC, and ADC that appear in real experiments.

Both anechoic and reverberant environments are considered in the experiments. Anechoic environments are simulated by choosing the reflection coefficients $\{\beta_i : i = 1, 2, \cdots, 6\}$ of all room boundaries equal to zero in

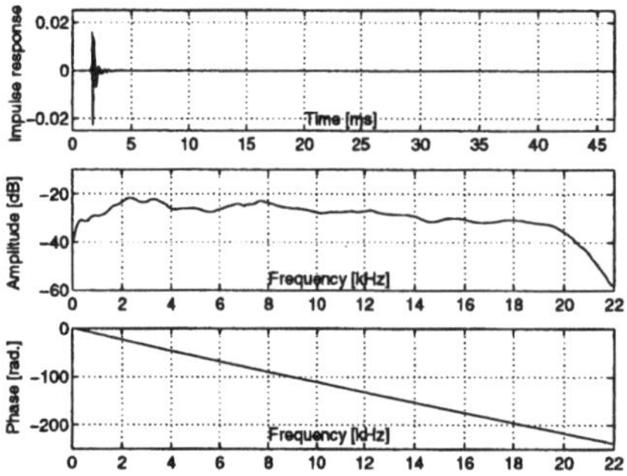

Figure 4.3: A calculated transfer function in an anechoic environment with $\beta_i = 0$.

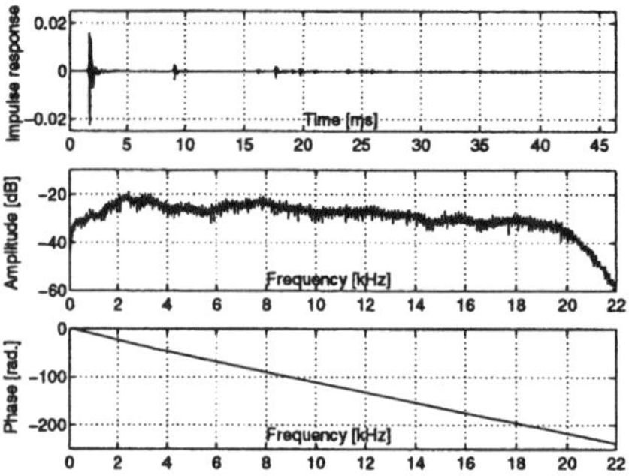

Figure 4.4: A calculated transfer function in a reverberant environment with $\beta_i = 0.5$.

the room simulation program. Reverberant environments are simulated by choosing these reflection coefficients $0 < |\beta_i| < 1$. In all experiments, the room dimensions in the x, y, and z directions are 6, 6, and 3 m, respectively. The origin of the coordinate system shown in Fig. 4.2 is taken at $x = 3, y = 3, z = 1$ m. Examples of the calculated transfer functions in an anechoic environment and a moderate reverberant environment are shown in Fig. 4.3 and Fig. 4.4, respectively.

4.2.4.2 Experiment Procedure

In all experiments discussed in this section, the microphones R_1 and R_2 were positioned on the x-axis at x = -10 and x = 10 cm, respectively. The desired response was chosen corresponding to an azimuth angle $\alpha = 60°$ and radial distance r = 1.5 m. The loudspeakers S_1 and S_2 were positioned at $\alpha = \mp 30°$, and r = 1 m. This simulates a virtual sound source in the horizontal plane at a wider azimuth angle as mentioned in Section 1.4.

To examine the performance of each of the above mentioned constraint methods under the same conditions, the system of constrained control equations for each method was solved in the frequency domain at each frequency to calculate the control filters $\underline{\mathbf{W}}_1(\omega)$ and $\underline{\mathbf{W}}_2(\omega)$. The control filters were implemented as FIR filters, each of 2048 taps. A third microphone was used to probe the sound field at different points in space. The room simulation program was used to calculate the transfer functions between the two loudspeakers S_1 and S_2, and the position of the third microphone. For simplicity of representing the results, derivative, difference, and weighted average constraints were considered only in the x-direction. Therefore, in the experiments presented below, the third microphone was moved on the x-axis from x = -20 to $x = 0$ cm in steps of 1 mm. Referring to the transfer functions from S_1 and S_2 to the third microphone as $\underline{\mathbf{C}}_{31}(\omega)$ and $\underline{\mathbf{C}}_{32}(\omega)$, respectively, the actual system response was then calculated at each position of the probing microphone using the relationship

$$\hat{\underline{\mathbf{H}}}_{31}(\omega) = \text{diag}\{\underline{\mathbf{C}}_{31}(\omega)\}\,\underline{\mathbf{W}}_1(\omega) + \text{diag}\{\underline{\mathbf{C}}_{32}(\omega)\}\,\underline{\mathbf{W}}_2(\omega). \tag{4.25}$$

This response was then transformed to the time domain (using an inverse FFT algorithm), resulting in $\hat{\underline{\mathbf{h}}}_{31}(n)$. It is worth mentioning at this point that direct solution of the system of control equations in the frequency domain may lead to non-causal filters $\underline{\mathbf{W}}_1(\omega)$ and $\underline{\mathbf{W}}_2(\omega)$. This causes a problem if the impulse response of those filters were to be directly calculated using the inverse Fourier transformation. However, frequency domain multiplication with $\underline{\mathbf{C}}_{31}(\omega)$ and $\underline{\mathbf{C}}_{32}(\omega)$, as in (4.25), introduces enough delay, resulting in a causal response $\hat{\underline{\mathbf{h}}}_{31}(n)$.

To examine the system performance in different frequency bands, the actual time response $\hat{\underline{\mathbf{h}}}_{31}(n)$ was filtered through a one octave constant-Q (percentage bandwidth) filterbank composed of eight filters covering the frequency range from 80 Hz to 20480 Hz, resulting in eight time responses $\{\hat{\underline{\mathbf{h}}}_{31i}(n) : i = 1, 2, \cdots, 8\}$. The desired responses $\underline{\mathbf{h}}_{11}(n)$ and $\underline{\mathbf{h}}_{21}(n)$ at microphones R_1 and R_2 were also filtered through the same filterbank, resulting in $\{\underline{\mathbf{h}}_{11i}(n) : i = 1, 2, \cdots, 8\}$ and $\{\underline{\mathbf{h}}_{21i}(n) : i = 1, 2, \cdots, 8\}$, respectively. In each frequency band i, the actual and desired responses were compared and an error measure for each desired response was calculated as

$$E_{ji} = 10 \log_{10}\left(\frac{\sum_n [\hat{h}_{31i}(n) - h_{j1i}(n)]^2}{\sum_n h_{j1i}^2(n)}\right), \qquad i = 1, 2, \cdots, 8, \; j = 1, 2. \tag{4.26}$$

The errors E_{1i} in all eight frequency bands were then plotted at each position of the probing microphone along the x-axis. Each experiment was then repeated exactly, except that the control filters were calculated using the system of unconstrained control equations given by (4.3). The errors resulting from the unconstrained filters were plotted on the same graphs for reference.

4.2.4.3 Derivative Approximations

An electro-acoustic transfer function between a receiver at (x_r, y_r, z_r) and a point monopole sound source at (x_s, y_s, z_s) in free field and at a far distance from the source may be approximated by

$$C_{rs}(\omega, x_r, y_r, z_r) = A_{rs}\, e^{-j\omega d_{rs}/c}, \tag{4.27}$$

where d_{rs} is the distance between the source and the receiver, c is the sound velocity in the medium, ω is the angular frequency of the sound and A_{rs} is a position dependent amplitude factor [10]. In the far field, and at small displacements from the receiver position, the amplitude factor A_{rs} may be considered distance independent. Using this approximation, the first derivative of $C_{rs}(\omega, x_r, y_r, z_r)$ with respect to x_r may be written as

$$\frac{\partial C_{rs}(\omega, x_r, y_r, z_r)}{\partial x_r} = \frac{j\,\omega\,(x_s - x_r)}{c\, d_{rs}}\, A_{rs}\, e^{-j\,\omega\, d_{rs}/c}. \tag{4.28}$$

Using transfer functions given by (4.27) and their first derivatives given by (4.28), it is shown in [10] by means of computer simulations that the first-order derivative constrained control filters designed using (4.8) result in wider *overall* zones of equalisation. Not shown in [10] are the performance in separate frequency bands and the validity of the approximation for practical electro-acoustical transfer functions.

As mentioned in Section 4.2.1, it becomes more difficult to calculate the analytic expressions for the transfer functions as the reverberation increases. In this section, a slight modification to the first derivative given by (4.28) that allows using derivative constrains in moderate reverberant conditions is proposed. The proposed approximation exploits the fact that a transfer function measured in a reverberant environment has an initial delay equals to that given by the exponential factor in (4.27). Since the derivative in (4.28) is derived using this delay factor only, a sound approximation in moderate reverberant environments is to apply the same delay constraint. Using such a delay constraint implies that the control filters force the direct sound to arrive at the same moment at points in the vicinity of the control microphone while no control effort is made to control the reflections. A control effort on the reflections may be added by deriving the derivative from a transfer function measured in the reverberant environment. Since the derivative in (4.28) is a high-pass

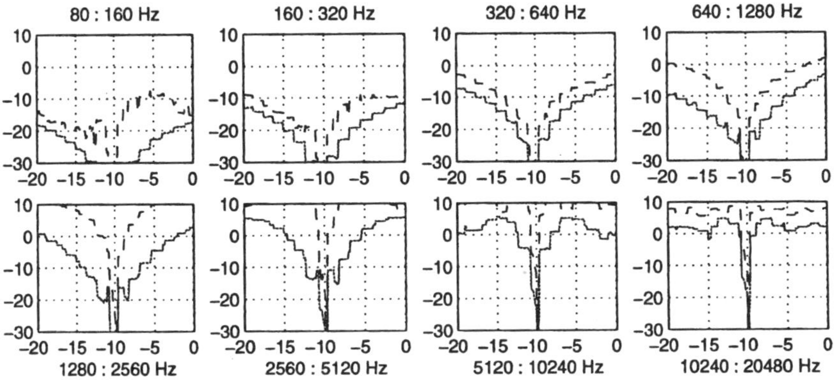

Figure 4.5: The errors $\{E_{1i} : i = 1, 2, \cdots, 8\}$ in a reverberant environment with reflection coefficients $\{\beta_i = 0.5 : i = 1, 2, \cdots, 6\}$ for derivative constrained filters (solid) and unconstrained filters (dashed). The horizontal axis represents the distance on the x-axis in cm from the control microphone R_1 located at x=-10cm. The vertical axis represents the error reduction in dB.

filtered version of the original transfer function, the derivative equation for a reverberant transfer function may be approximated by

$$\frac{\partial C_{rs}(\omega, x_r, y_r, z_r)}{\partial x_r} = G_{HP}(\omega)\ C_{rs}(\omega, x_r, y_r, z_r), \qquad (4.29)$$

where $G_{HP}(\omega)$ is a high-pass filter function. In severe reverberant conditions, and at distances further than the reverberation distance (see Section 2.1.6.3), the power of the diffuse sound field may be larger than that of the direct sound, and the above approximation may fail.

The performance of the derivative constrained filters $\underline{\mathbf{W}}_1$ and $\underline{\mathbf{W}}_2$ shown in Fig. 4.2 when the derivatives are calculated using (4.29) is shown in Fig. 4.5. Figure 4.5 shows the error at the probing microphone with respect to the desired response at R_1 (located at x = -10 cm) in each octave band as mentioned in Section 4.2.4.2. This result was obtained using transfer functions calculated in a room with reflection coefficients $\{\beta_i = 0.5 : i = 1, 2, \cdots, 6\}$. For 10 dB equalisation level, the constrained filters achieve wider equalisation zones than the unconstrained filters in all but the highest two octaves. The 20 dB equalisation zone is approximately 20 cm in diameter in the lowest octave, while the level and the size of the zones decrease as the frequency increases.

The performance of the derivative constrained filters can be understood from the physical nature of the constraint. This is best explained using a single channel system. In this case, the system of derivative constrained control equations is given by

$$\begin{aligned} C_{11}(\omega)\, W_{11}(\omega) &= H_{11}(\omega), \\ G_{HP}(\omega)\, C_{11}(\omega)\, W_{11}(\omega) &= 0. \end{aligned} \tag{4.30}$$

Since the system of equations is overdetermined, there exists no exact solution. A general LMS solution of such an overdetermined system is $\mathbf{W} = (\mathbf{C}^H\mathbf{C})^{-1}\mathbf{C}^H\mathbf{H}$, where $\cdot^H$ denotes the complex conjugate transpose (Hermitian). The LMS solution for (4.30) is, therefore,

$$W_{11}(\omega) = \frac{1}{1 + |G_{HP}(\omega)|^2}\, C_{11}^{-1}(\omega)\, H_{11}(\omega). \tag{4.31}$$

At very low frequencies, $|G_{HP}(\omega)|^2$ is very small, and the solution to the filter is $W_{11}(\omega) \approx C_{11}^{-1}(\omega)\, H_{11}(\omega)$, which is very close to the exact solution. As the frequency increases, the term $|G_{HP}(\omega)|^2$ increases and the filter drifts from the exact solution. The filter is, therefore, designed to increasingly ignore more details in the acoustic transfer function $C_{11}(\omega)$ as the frequency increases. This makes the filter a good approximation to the solution required at points in the vicinity of the original control point. The area in which the filter is considered a valid solution decreases with increasing frequency. In fact, (4.31) suggests that the general solution for the derivative constrained filter is a low-pass filtered version of the exact solution. The idea of designing the control filters to increasingly ignore details as the frequency increases is further pursued in Section 4.3.

4.2.4.4 Microphone Clusters

Figure 4.6 shows the error criteria for the first microphone $\{E_{1i} : i = 1, 2, \cdots, 8\}$ when difference constraints are used to calculate the control filters. This result was obtained by using two extra microphones positioned at $x = -15$ cm and $x = -5$ cm in addition to the original microphone R_1 at $x = -10$ cm. In the room, the three microphones were positioned at $(x = 2.85m, y = 3m, z = 1m)$, $(x = 2.95m, y = 3m, z = 1m)$, and $(x = 2.90m, y = 3m, z = 1m)$, respectively. Transfer functions calculated for a room with reflection coefficients $\{\beta_i = 0.5 : i = 1, 2, \cdots, 6\}$

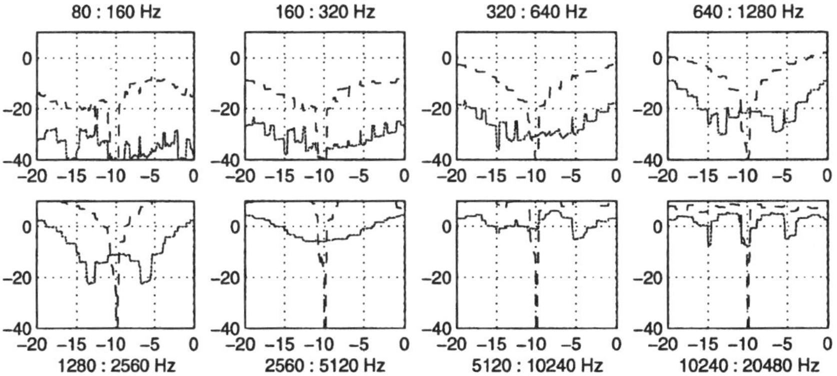

Figure 4.6: The same as Fig. 4.5 but with the solid curve obtained for difference constrained filters. Two extra microphones positioned at x = - 15 and x = - 5 cm were used.

have been used in this simulation. It can be seen from this result that microphone clusters are very effective at low frequencies. The 30 dB zone of equalisation is wider than 20 cm in the lowest octave. The 20 dB zone of equalisation of 20 cm diameter is extended to cover the frequency range up to 640 Hz, while the same zone was limited to 160 Hz for derivative constrained filters. The single zone of equalisation becomes narrower and shallower as the frequency increases. Above a certain frequency (which depends on the separation between the microphones), the single large zone of equalisation splits into separate zones around the individual microphones. This effect is very clear in the highest octave band where very narrow and shallow zones of equalisation are created around the three microphones. Decreasing the distance between the microphones extends the zones to higher frequencies while reducing the diameter of the single large zone. Using more microphones (say at x = -15, x = -12.5, x = -7.5, and x = -5 cm) extends the frequency range while keeping the width of the single zone unchanged. However, the equalisation level may decrease as the sound field is controlled at more and more microphones using only two loudspeakers.

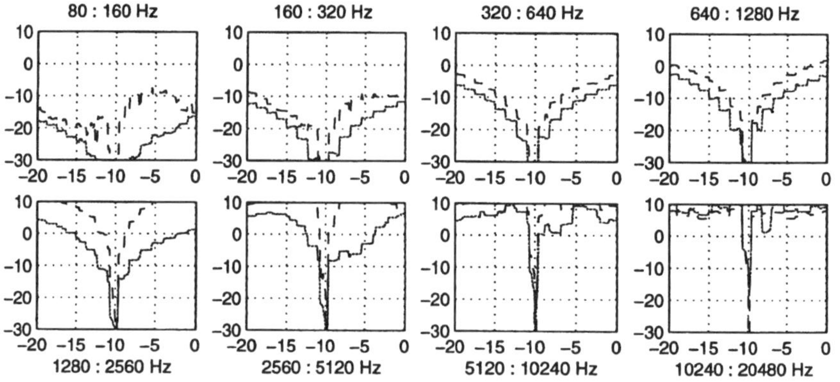

Figure 4.7: The same as Fig. 4.5 but with the solid curve obtained for weighted average constrained filters. Two extra transfer functions from each loudspeaker to $x = -12$ and $x = -8$ cm were used to calculate the weighted average.

4.2.4.5 Spatial High-Pass Filters

As mentioned in Section 4.2.3, the complexity of the system of difference constrained control equations given by (4.17) may be significantly reduced by employing a single constraint derived by spatially filtering all transfer functions measured at each microphone cluster. Extensive simulations show that only spatial high-pass filters are capable of widening the zones of equalisation. This is also suggested by (4.22), since the sign of the centre coefficient is opposite to the other coefficients. Such a spatial high-pass filter rejects common features among the transfer functions in the cluster while retains fast changing features. Since fast changing features increasingly appear as the frequency increases, a spatial high-pass filter may be seen to reject the low frequencies and retain the high frequencies of the transfer function measured at the control point in the centre of the cluster. This is similar to the action of the high-pass filter $G_{HP}(\omega)$ in (4.29), which explains the reason why the spatial filter must be of a high-pass nature. The weighted average constrained control filters, therefore, approximate the derivative constrained filters only if the weighted average is of the high-pass type. The more transfer functions are used to calculate the weighted average, the better the approximation becomes.

An example of the performance of weighted average constrained filters is shown in Fig. 4.7. Transfer functions calculated for a room with reflection coefficients $\{\beta_i = 0.5 : i = 1, 2, \cdots, 6\}$ have been used in this simulation. The weighted average transfer functions $\check{C}_{1l}$ in (4.24) were obtained by using the following high-pass spatial filter

$$\check{C}_{1l} = \frac{2}{6} C_{1l} - \frac{4}{6} C_{2l} + \frac{1}{6} C_{3l}, \qquad l = 1, 2, \tag{4.32}$$

where C_{1l}, C_{2l} and C_{3l} are the transfer functions between microphones positioned at x = -12, x = -10, and x = -8 cm and the lth loudspeaker, respectively. This result shows that simple spatial high-pass filters are capable of approximating the derivative constraint and, therefore, result in wider equalisation zones. Similar properties of the zones of equalisation in the different frequency bands may also be observed from this result as in the cases of derivative and difference constraints.

4.3 Multiresolution Control Filters

Assuming that there exists an exact solution for the matrix of control filters $\mathbf{W}(\omega)$ in the generalised multichannel audio reproduction system shown in Fig. 3.1, this solution is given by

$$\mathbf{W}(\omega) = \mathbf{C}^{-1}(\omega)\mathbf{H}(\omega). \tag{4.33}$$

The factor $\mathbf{C}^{-1}(\omega)$ exactly inverts the electro-acoustic channels between the reproduction loudspeakers and the listeners' ears, and the factor $\mathbf{H}(\omega)$ adds the desired response as mentioned before. The main problem with this exact solution is that it is valid only at the listeners' ears, the points at which the transfer functions $\mathbf{C}(\omega)$ and $\mathbf{H}(\omega)$ have been previously measured. When a listener moves slightly, both $\mathbf{C}(\omega)$ and $\mathbf{H}(\omega)$ change, and the control filters $\mathbf{W}(\omega)$ become in error, which is a problem that is inherent to the exact solution. Due to the short wavelengths at high frequency, two impulse responses measured at adjacent points in a room increasingly differ from each other as the frequency increases. The same short wavelength also makes head-related transfer functions increasingly differ from those measured at adjacent azimuth or elevation angles as the frequency increases. These physical properties cause the errors introduced in $\mathbf{W}(\omega)$ due to a listener movement to increase with increasing frequency.

One approach to improving the robustness of $\mathbf{W}(\omega)$ against small listener movements is to design the filters to be an average solution that is valid in an area in space around the listeners' ears. This is in contrast to filters that implement the exact solution, which is valid only at the listeners' eardrums. Based on the above mentioned physical observations, the deviation from the exact solution should also increase with increasing frequency to cope with the error increase. This is also in full agreement with the physical interpretation of the constrained filters discussed in Section 4.2.

Many authors have reported obtaining robust systems using control filters with average solution at high frequencies. It is suggested in [46] that using transfer functions that are free from details of specific heads to design the control filters results in filters that are tolerant to variations in head shape and position. It is also suggested in [46] that if robust filters are required, only the envelope at high frequencies and not the detailed solution should be employed. In other work [92], one-third octave constant-Q smoothing filters were used to smooth the HRTF data prior to using them in calculating the cross-talk cancellation and the binaural synthesis filters. The same constant-Q smoothing was also used in [72] for implementing cross-talk cancellers. Furthermore, the cross-talk cancellation function employed in [72] was band-limited to 6 kHz, while only the correct sound power is delivered to the ears at higher frequencies[3]. In yet another application [76, 77], segmented adaptive control filters were used, where the control filters were divided into segments with lower frequency resolution as the frequency increases.

In all the above mentioned examples, the main goal was to average the sharp features that occur at high frequencies from the HRTF data to avoid a solution that deals specifically with those features. Since those sharp and fast changing features increase with increasing frequency, it is essential that the width of the averaging window also increases with frequency. This has already been applied in the constant-Q averaging [92] and the segmented filters [76, 77] mentioned above. The resulting control filters are of non-uniform frequency resolution (multiresolution filters) with decreasing frequency resolution with increasing frequency, therefore,

[3]Band-limiting the cross-talk canceller was first suggested in [47]

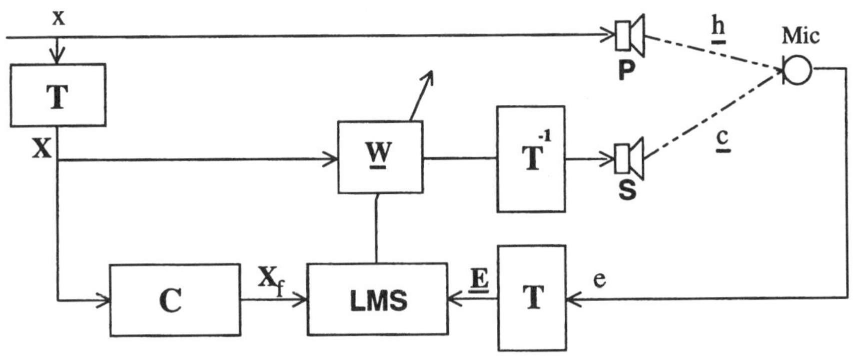

Figure 4.8: A multiresolution adaptive filter.

averaging wider frequency bands at shorter wavelengths. This approach, in addition to leading to an average solution at high frequencies, relaxes the problem of inverting deep notches in the cross-talk canceller.

On the other hand, the multiresolution filters may sound objectionable in some applications such as virtual source synthesis, since sound localisation cues for elevation and cone of confusion resolution are high frequency features that are to be averaged. The effects of this averaging on the perceived source location can only be studied using psychoacoustical experiments. However, the use of multiresolution filters is justified by the fact that the human auditory system performs a similar non-uniform spectral analysis and, therefore, has similar inaccuracies as mentioned in Section 2.1.8. Furthermore, human localisation resolution that depends on high frequency cues (elevation resolution for instance) are rather poor compared to those based on binaural cues [30]. Since a multiresolution filter is still delivering the correct interaural time delay envelope at high frequencies and correct interaural intensity difference, the effects on localisation accuracy may be reduced. This, however, must be confirmed by psychoacoustical experiments.

In conventional 3D sound systems, the control filters are usually designed off-line, and the multiresolution control filters may be calculated by processing the HRTF data using constant-Q smoothing filters. In adaptive 3D sound systems, this frequency-dependent averaging must be performed in real-time. When the adaptive filters are implemented in the frequency domain, this reduces to forcing the adaptive filter to the required resolution by employing a non-uniform, constant-Q-like transformation in place of the constant bandwidth FFT, as shown in Fig. 4.8. The problem of designing the multiresolution filters then reduces to designing the multiresolution transformation **T** that replaces the FFT in conventional frequency domain adaptive filters. Since computational saving is one of the main reasons for implementing the filters in the frequency domain as mentioned in Section 3.4.2, the transformation **T** must be calculated as fast as the usual FFT to maintain this computational advantage. Chapter 5 is devoted to developing such fast multiresolution transformations, where it is shown that a fast multiresolution transformation may be obtained by calculating the FFT of non-uniformly sampled signals [66, 70]. Improving the robustness of 3D sound systems by employing multiresolution control filters is shown by computer simulations in [68], and repeated in Section 4.3.1 for reference. Most audio signal processing applications may also benefit from this non-uniform spectral resolution that resembles the processing of the human ears. An example of improving the performance of phase vocoder systems has also been presented in [69].

4.3.1 Performance of Multiresolution Filters

In this section, it is shown, using a computer experiment, that the robustness of 3D sound systems can be improved by using control filters having non-uniform frequency resolution as mentioned in Section 4.3. The experiment shown in Fig. 4.9 implements a one-point noise canceller, which may also be seen as a simplified version of a virtual sound source generator at one ear only. A white noise signal x is filtered through an acoustic impulse response $\underline{\mathbf{h}}(n)$ simulating the sound pressure due to a physical loudspeaker P at one eardrum of the listener. The filter $\underline{\mathbf{w}}$ is calculated to minimise the sound pressure at MIC through loudspeaker S and the acoustic transfer function $\underline{\mathbf{c}}(n)$. The MIC output is then fed to a cochlea model (gammatone filterbank [117]) consisting of 64 channels and covering the frequency range from 20 Hz to 20 kHz. The output of each fourth channel of these 64 channels are half-wave rectified, low-pass filtered and plotted against time for inspection. These plots represent the probability of firing along the auditory nerve due to the sound signal received at the listener's eardrum represented here by the microphone MIC.

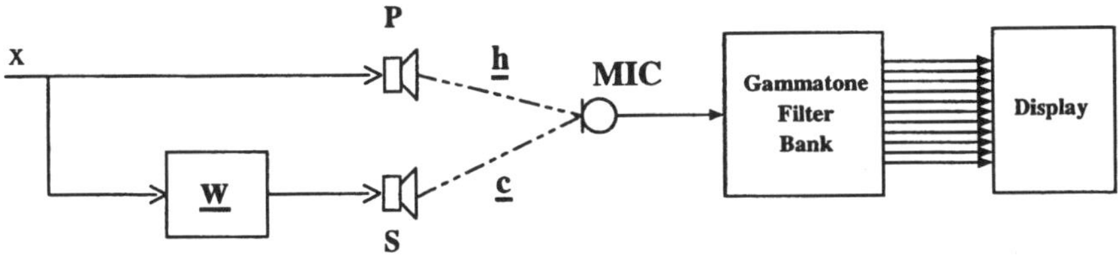

Figure 4.9: Experiment set-up.

The optimum solution for the filter $\underline{\mathbf{w}}$ in this case is given in the frequency domain by

$$\underline{\mathbf{W}}(\omega) = \text{diag}\{\underline{\mathbf{C}}^{-1}(\omega)\}\, \underline{\mathbf{H}}(\omega), \tag{4.34}$$

where $\underline{\mathbf{C}}(\omega)$ and $\underline{\mathbf{H}}(\omega)$ represent the spectra of the corresponding acoustic impulse responses at ω. From this equation, it is clear that the filter length must be large enough to accommodate this solution. The degree of attenuation in sound pressure at MIC depends on the length of the filter $\underline{\mathbf{w}}$. A longer filter achieves higher attenuation and, therefore, better emulation of the virtual source.

Using the FFT algorithm to calculate $\underline{\mathbf{C}}(\omega)$ and $\underline{\mathbf{H}}(\omega)$ from $\underline{\mathbf{c}}(n)$ and $\underline{\mathbf{h}}(n)$ produces a constant resolution filter $\underline{\mathbf{W}}(\omega)$. On the other hand, using a multiresolution transformation to obtain $\underline{\mathbf{C}}$ and $\underline{\mathbf{H}}$ produces a multiresolution solution for $\underline{\mathbf{W}}$. To show the performance improvement achieved by using multiresolution filters, the following experiment was performed:

Four acoustic impulse responses, each of 512 samples, were calculated using the room simulation program *Room Impulse Response 2.1* [71]. A room having a reverberation time of 0.3238 second, and dimensions of 5 m, 4 m, and 3 m in the x, y, and z directions, respectively, has been used. The four transfer functions are as follows:

- $\underline{\mathbf{h}}_0$: from loudspeaker P at (2, 3, 1.5 m) to MIC position at (4, 3, 1.5 m)
- $\underline{\mathbf{c}}_0$: from loudspeaker S at (4, 2, 1.5 m) to MIC position at (4, 3, 1.5 m)
- $\underline{\mathbf{h}}_1$: from loudspeaker P at (2, 3, 1.5 m) to MIC position at (4, 3, 1.52 m)

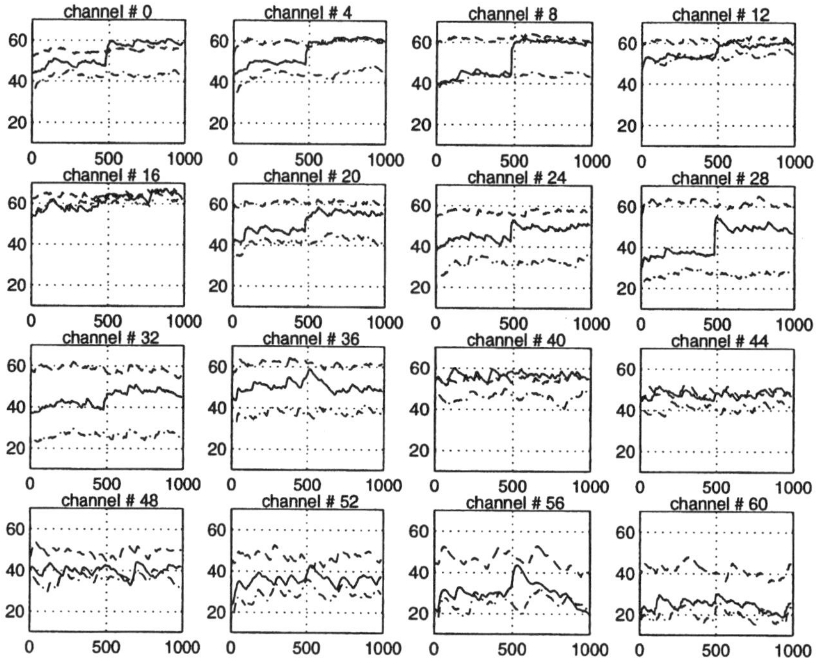

Figure 4.10: Output of the cochlea model: channel # 0 corresponds to the highest frequency band. The horizontal axis represents time in samples and the vertical axis is the sound pressure level (SPL) in dB. (– –) is the SPL before cancellation, (—) and (–·–) are the SPL after cancellation using a uniform and non-uniform resolution filters, respectively.

- $\underline{c}_1$: from loudspeaker S at (4, 2, 1.5 m) to MIC position at (4, 3, 1.52 m)

$\underline{\mathbf{h}}_1$ and $\underline{c}_1$ represent moving MIC 2 cm vertically in space from its initial position. The filter $\underline{\mathbf{W}}$ of 1024 coefficients was calculated in two different ways: using uniform and non-uniform spectral resolutions. Only the initial MIC position has been used in calculating $\underline{\mathbf{W}}$ such that $\underline{\mathbf{W}} = \text{diag}\{\underline{\mathbf{C}}_0^{-1}\}\,\underline{\mathbf{H}}_0$. The simulation shown in Fig. 4.9 was repeated twice: once using the constant resolution filter and the second time using the multiresolution one. At the middle of the simulation, MIC is moved 2 cm from its initial position (the simulation switched to $\underline{\mathbf{h}}_1$ and $\underline{c}_1$ in place of $\underline{\mathbf{h}}_0$ and $\underline{c}_0$). The output of this experiment is shown in Fig. 4.10, where the horizontal axis represents time in samples and the vertical axis is the sound pressure level (SPL) in dB. The simulation represents moving the microphone 2 cm vertically after 500 samples. The reduction achieved by a robust filter should not be affected by this sudden movement. From this experiment, following may be concluded:

- The constant resolution solution suffers from the microphone movement only at high frequencies, where the movement is comparable to the wavelength. This is consistent with the motivation of using multiresolution approach as discussed in section 4.3.
- The multiresolution filter results in a better attenuation in almost all frequency bands. At low frequencies, higher resolution is used and, therefore, a more accurate solution is obtained. At high frequencies, coarser resolution is used, which prevents dealing specifically with sharp features and matches the resolution of the cochlea model.
- Spatial averaging at high frequencies is clear when using coarse resolution as discussed in section 4.3. This averaging effect makes the solution more robust to movements.
- The constant resolution filter could not achieve any attenuation at some frequencies (such as channels 40 and 44) even before moving the microphone, while the multiresolution filter could. This may occur if the spectrum of $\underline{c}$ contains a deep notch at these frequencies, causing inversion problems. Since coarser resolution is equivalent to spectral smoothing, this produces shallower notches and a better solution for $\underline{\mathbf{W}}$ may be obtained.

4.4 Underdetermined Systems and Exact Solutions

Another important property of the zones of equalisation, besides their width, is the equalisation level. One expects that using more degrees of freedom in the generalised multichannel audio reproduction system shown in Fig. 3.1 results in a higher equalisation level due to improved controllability. More degrees of freedom may be employed by increasing the number of taps of the control filters. Better results may be obtained by dividing the number of taps into more filters driving more loudspeakers. As mentioned in Section 3.2, when the number of reproduction loudspeakers L is greater than the number of control microphones M, the system of control equations is underdetermined. There are less equations than unknowns, and there exists an infinite number of solutions. However, it is shown in [102, 103] that when $L = M + 1$, there exists a unique and exact (opposed to LMS) solution for the matrix of control filters $\mathbf{W}(\omega)$, which reduces the error exactly to zero provided that

- All impulse responses $\{\underline{\mathbf{c}}_{ml} : l = 1, 2, \cdot, L\}$ from all the loudspeakers to the mth microphone do not contain any common zeros.
- The number of FIR coefficients of the control filters N_w is constrained to be less than the number of FIR coefficients N_c representing $\underline{\mathbf{c}}_{ml}$.

This approach, referred to as the Multiple-input\output INverse Theorem (MINT), is further extended in [107] for $L > M$ not constrained to $L = M + 1$. Furthermore, it is shown in [107] that with the correct choice of both modelling delay and the number of coefficients of the control filters, the design of the filters using MINT or MELMS (see Chapter 3) should in principle yield identical results. Therefore, it is possible to implement the system using adaptive filters, the optimum solutions of which are the exact solutions that reduce the error exactly to zero by simply increasing the number of reproduction loudspeakers and fulfilling the delay and number of coefficients constraints. Provided that the loudspeakers are well-positioned in the listening space (see Section 4.5), the underdetermined system is expected to provide better equalisation level than the fullydetermined system.

The correct modelling delay and the number of coefficients for the adaptive filters to exactly model the solution have also been evaluated in [107]. As mentioned in Section 3.3.4, the target transfer functions $\underline{\mathbf{H}}_{mk}(\omega)$ must contain sufficient delay for a causal and stable solution to exist, which is the same constraint on the modelling delay. Assume that the shortest pure delay in the impulse responses $\{\underline{\mathbf{c}}_{ml} : l = 1, 2, \cdots, L\}$ between any of the reproduction loudspeakers and the mth microphone is Δ_m^{min} samples, and all $\{\underline{\mathbf{c}}_{ml} : l = 1, 2, \cdots, L, m = 1, 2, \cdots, M\}$ can be accurately modelled by FIR filters, the longest of which contains no non-zero

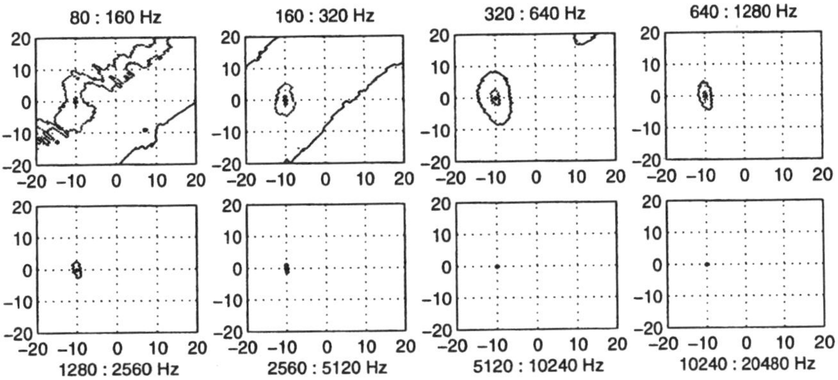

Figure 4.11: Equal contours of the error criteria $\{E_{1i} : i = 1, 2, \cdots, 8\}$ at equalisation levels of 0, -10, and -20 dB in a room with $\beta_i = 0.5$, obtained using two loudspeakers positioned at $\alpha = \pm 45°$. Both axes represent the distance in cm from the target point at $x = -10$ cm and $y = 0$.

coefficients beyond $\Delta_m^{max} - 1$ samples. If the adaptive filters are FIR filters of equal length N_w, then the delay Δ_m in any of the target transfer functions $\{\underline{\mathbf{H}}_{mk} : k = 1, 2, \cdots, K\}$ must be in the range [107]

$$\Delta_m^{min} \leq \Delta_m < \Delta_m^{max} + N_w - 1. \tag{4.35}$$

The number of coefficients of each of the control filters for the exact solution to exist must be

$$N_w = \frac{\left[\sum_{m=1}^{M} \left(\Delta_m^{max} - \Delta_m^{min}\right)\right] - M}{L - M}. \tag{4.36}$$

To examine the improvement brought about by using more reproduction loudspeakers, the experiment described in Section 4.2.4.2 has been repeated twice. Once with two loudspeakers positioned at $\alpha = \pm 45°$ and a radial distance $r = 100$ cm from the centre of the listener's head and once with three loudspeakers positioned at $\alpha_1 = -45°$, $\alpha_2 = +45°$, $\alpha_3 = 180°$, and $r = 100$ cm. The error criteria (4.26) were calculated at discrete points in the range from $x = y =$ -20 cm to $x = y = 20$ cm with 2 mm resolution in the horizontal plane $y = 100$ cm containing the microphones and the loudspeakers. Plots of the equal contours

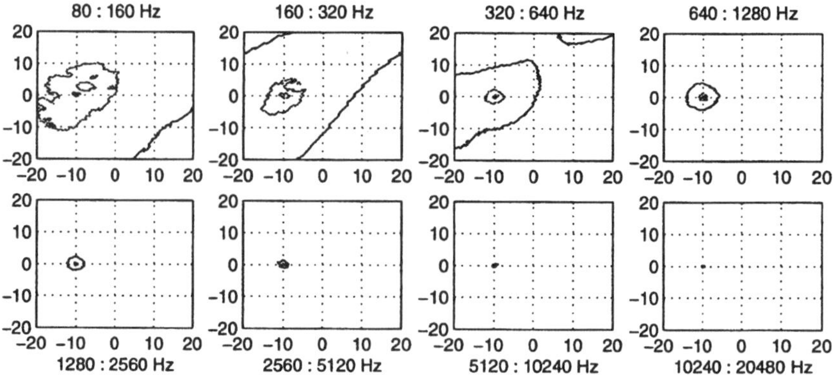

Figure 4.12: The same as Fig. 4.11 obtained using three loudspeakers positioned at $\alpha_1 = -45°$, $\alpha_2 = +45°$, and $\alpha_3 = 180°$.

of $\{E_{1i} : i = 1, 2, \cdots, 8\}$ at equalisation levels of 0, -10, and -20 dB are shown in Fig. 4.11 and Fig. 4.12 for the two and three loudspeakers, respectively. Similar computer simulations in [107] also show that choosing $L > M$, with the correct delay and number of coefficients, not only decreases the error at the control points considerably, but also results in slightly wider zones of equalisation at mid-band frequencies. Cross-talk cancellation experiments at two microphones conducted in [139] also show higher equalisation levels for three reproduction loudspeakers.

A more important effect of the underdetermined system is the shape of the zones of equalisation. The three-loudspeaker system results in zones that are almost round, therefore, allowing the listener to move his head equally in all directions. This is in contrast to the two-loudspeaker system, which allows the listener to move his head to the front and back more than to the right and left as shown in Fig. 4.11 and Fig. 4.12. This, however, is not the case for arbitrary positions of the loudspeakers. The positions of the loudspeakers in the three loudspeakers experiment has been optimally chosen (see Section 4.5).

Underdetermined systems also allow dividing the control effort on more filters and more loudspeakers. It was noticed from the above mentioned simulations that the values of the filters' coefficients in the three-loudspeaker system are smaller than those for the two-loudspeaker system at all frequencies. Therefore, underdetermined systems are less likely to overload the loudspeakers and cause nonlinearities.

4.5 Reproduction Loudspeakers Positioning

It is shown in Section 3.3.3 that the positions of the reproduction loudspeakers relative to the control microphones have considerable effect on the convergence speed of the adaptive filters. This same relative positions also have strong influence on the sensitivity of the filters to listeners' movements. This can be explained by examining the control equations for the kth input signal

$$\mathbf{C}(\omega)\underline{\mathbf{W}}_k(\omega) = \underline{\mathbf{H}}_k(\omega), \tag{4.37}$$

where $\underline{\mathbf{W}}_k(\omega)$ and $\underline{\mathbf{H}}_k(\omega)$ are the columns in $\mathbf{W}(\omega)$ and $\mathbf{H}(\omega)$ given in (4.3), corresponding to the kth input signal. From error estimates in numerical analysis, it is known that in a system of linear algebraic equations such as (4.37), a small change in $\mathbf{C}(\omega)$ causes a large change in the solution $\underline{\mathbf{W}}_k(\omega)$ when $\mathbf{C}(\omega)$ is ill-conditioned [5, 10, 36]. Considering small listener movements as small changes in $\mathbf{C}(\omega)$, the difference between the required solution for $\underline{\mathbf{W}}_k(\omega)$ before and after a movement is large if $\mathbf{C}(\omega)$ is ill-conditioned. Small movement lead to large errors in the filters, therefore, a system with an ill-conditioned matrix $\mathbf{C}(\omega)$ is more sensitive to listeners' movements.

Since the condition of the matrix $\mathbf{C}(\omega)$ is dependent on the positions of the reproduction loudspeakers relative to the microphones, it is important to choose those positions so that the matrix $\mathbf{C}(\omega)$ is well-conditioned at all frequencies. This may be done by examining the condition number of $\mathbf{C}(\omega)$ at each frequency in the frequency range of interest. The 2-norm condition number of a matrix $\mathbf{C}(\omega)$ is defined as the ratio between its largest and smallest singular values[4] [10]

$$\kappa_2\{\mathbf{C}(\omega)\} = \frac{\sigma_{max}\{\mathbf{C}(\omega)\}}{\sigma_{min}\{\mathbf{C}(\omega)\}}. \tag{4.38}$$

A well-conditioned matrix has a condition number close to one, while an ill-conditioned matrix has a large condition number. Ill-conditioned situations, such as those described in Section 3.3.3, reduce the rank of the matrix $\mathbf{C}(\omega)$ and consequently increase its condition number making it more sensitive to listener's movements.

[4] The singular values of a matrix $\mathbf{A}$ are equal to the square roots of the eigenvalues of $\mathbf{A}\mathbf{A}^H$.

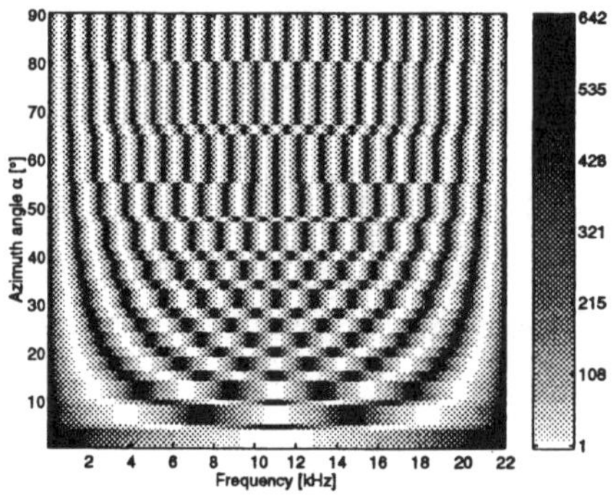

Figure 4.13: The condition number of $\mathbf{C}(\omega)$ in an anechoic room. $d_m = 18$ cm, and $r = 100$ cm.

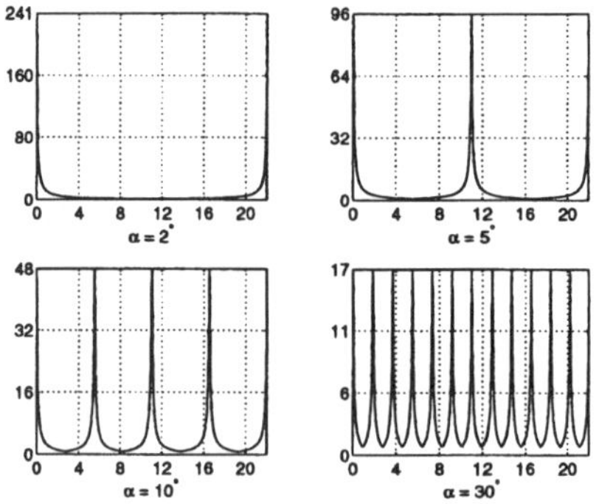

Figure 4.14: The condition number of $\mathbf{C}(\omega)$ at discrete angles from those shown in Fig. 4.13

4.5.1 Optimum Stereo Set-Up

A case of special interest in audio reproduction is the stereo system, where two reproduction loudspeakers are used. Optimum positioning of those two loudspeakers may be investigated by examining the condition number (4.38) for different positions of the loudspeakers at each frequency ω. The experimental set-up shown in Fig. 4.2 has been used to perform this investigation. In the following experiments, the azimuth angle α was changed from $1°$ to $90°$ in one degree steps, while the radial distance r is kept constant. In other words, the two loudspeakers were moved on a circle of radius one meter, the centre of which is the centre of the listener's head. In each position, the room simulation program *Room Impulse Response 2.1* [71] was used to calculate the four electro-acoustic transfer functions between the loudspeakers and the microphones. Each transfer function was calculated as an FIR filter of 2048 taps at 44.1 kHz sampling frequency. The FFT algorithm was used to calculate the discrete frequency response of the four transfer functions at each position and (4.38) was used to calculate the condition number of $\mathbf{C}(\omega)$ at each discrete frequency ω. The above procedure was repeated for anechoic and reverberant conditions of different reverberation characteristics.

The condition number of $\mathbf{C}(\omega)$ in an anechoic chamber is shown in a grey-scale graph against the frequency in Fig. 4.13. In this experiment, the two microphones were positioned at $x = -9cm$ and $x = 9cm$. The radial distance between the loudspeakers and the centre of the listener's head was $r = 100$ cm. Figure 4.14 shows the condition number at selected values of α from those shown in Fig. 4.13 in two-dimensional graphs. The horizontal axis in Fig. 4.14 represents the frequency and the vertical axis represents the condition number. Since the condition number of $\mathbf{C}(\omega)$ in an anechoic environment is dependent only on the relative positions of the loudspeakers and the microphones, different results are expected for different r. Figures 4.15 and 4.16 show the condition number of $\mathbf{C}(\omega)$ for the cases $r = 50$ cm, and $r = 150$ cm, respectively. The result shown in Fig. 4.16 was obtained by moving the two microphones from their symmetrical positions to $x = -7cm$ and $x = 11cm$.

Figure 4.14 suggests a physical interpretation for the relation between the mathematical concept of condition number and the robustness against listeners' movements. At frequencies where the condition number is large, the matrix $\mathbf{C}$ is almost singular, consequently, the responses of the filters have large peaks at those frequencies. Small movements may result in great changes in those peaks, making the filters very sensitive to movements. The following may also be deduced from those graphs:

- At small α, the matrix is ill-conditioned at low frequencies and well-conditioned at mid-band frequencies. The ill-condition at low frequencies is explained by the maximum similarity between the four transfer functions for small α.
- As α increases, the large condition number at low frequencies decreases and shrinks to cover a narrower frequency band. At the same time, spatial aliasing starts to appear. The amount of aliasing increases with increasing α.
- As the number of peaks due to spatial aliasing increases, the value of the condition number is equally divided into the new replicas as seen from Fig. 4.14. The matrix is well-conditioned in small frequency bands between the peaks.
- Although the traces do not change with changing the range r, the maximum value of the condition number increases with increasing r. This can be seen from the scales in the three grey-scale graphs. This is expected since as r increases, the similarity between the four transfer functions increases.

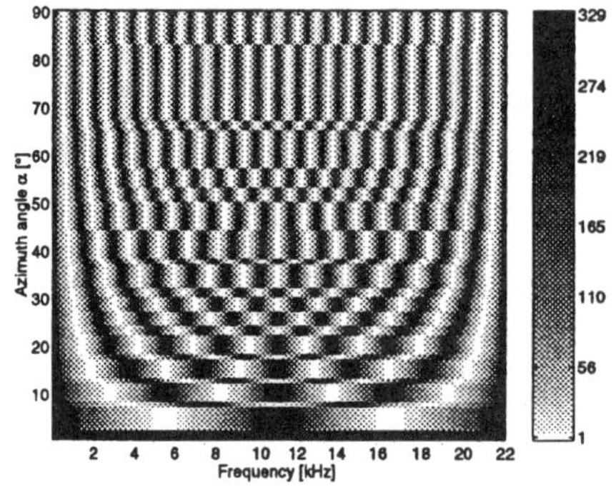

Figure 4.15: The condition number of $\mathbf{C}(\omega)$ in an anechoic room. $d_m = 18$ cm, and $r = 50$ cm.

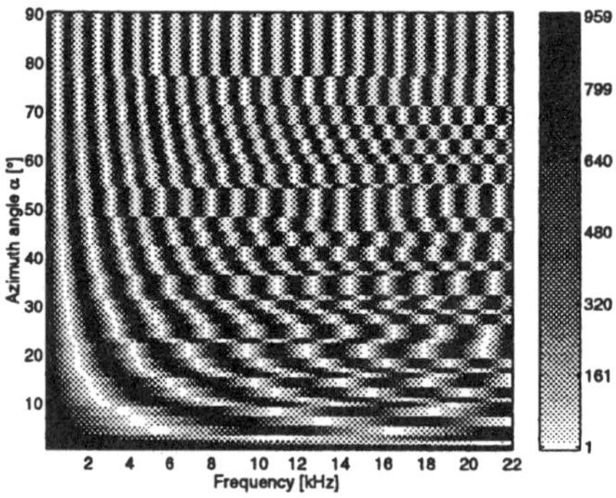

Figure 4.16: The condition number of $\mathbf{C}(\omega)$ in an anechoic room. $d_m = 18$ cm, and $r = 150$ cm.

It may be concluded from the above observations that a small α creates a wide frequency band at mid-band frequencies where the condition number is very low, while a large α does better at low frequencies. Although several authors [23, 91, 111] have suggested using closely spaced loudspeakers in stereo reproduction systems to widen the equalisation zones in the middle-band frequencies, this compromises the performance at low frequencies considerably. This problem has also been recognised in [91]. For closely spaced loudspeakers, and at low frequencies, the matrix $\mathbf{C}(\omega)$ is almost singular. From (4.37), the amplitude responses of the filters $\underline{\mathbf{W}}_k(\omega) = \mathbf{C}^{-1}(\omega)\underline{\mathbf{H}}_k(\omega)$ are, therefore, very large at low frequencies. The filters tend to amplify the low frequencies, which leads to saturating the audio amplifiers driving the loudspeakers or damaging the loudspeakers. A possible solution for this problem is to divide the audible frequency range into two or more bands. The low frequency band is reproduced through widely spaced loudspeakers while the high frequency band through closely spaced ones. The two closely spaced loudspeakers may be small tweeters since they reproduce high frequencies only, while the widely spaced ones must be large loudspeakers (subwoofers) for low frequency reproduction. Although this arrangement uses four loudspeakers, it may still be considered as a two-loudspeaker system, since each tweeter and subwoofer may be mounted on one enclosure, as is usually the case in closed or vented box loudspeaker systems. The two enclosures are then placed with the tweeters close to each other while the subwoofers are widely spaced. For most loudspeakers, this will mean

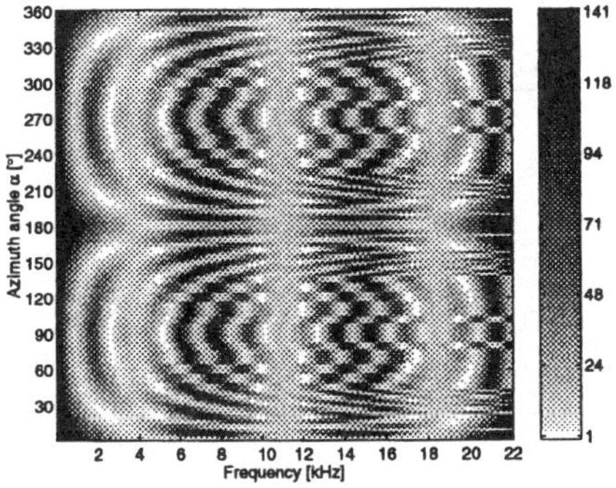

Figure 4.17: The condition number of $\mathbf{C}(\omega)$ in an anechoic room for three loudspeakers. Two loudspeakers are placed at azimuth angles $\alpha = \pm 5°$ while the position of the third is variable. $d_m = 18$ cm, and $r = 100$ cm.

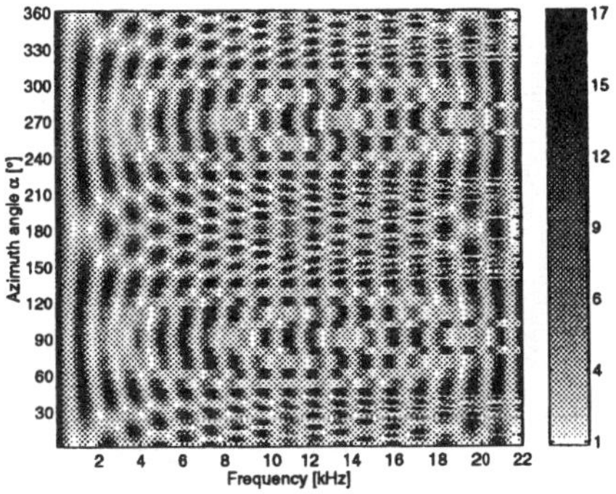

Figure 4.18: The condition number of $\mathbf{C}(\omega)$ in an anechoic room for three loudspeakers. Two loudspeakers are placed at azimuth angles $\alpha = \pm 45°$ while the position of the third is variable. $d_m = 18$ cm, and $r = 100$ cm.

placing the boxes horizontally in front of the listener instead of the conventional vertical position.

4.5.2 Three-Loudspeaker Systems

Another approach to improve the system robustness is to increase the number of reproduction loudspeakers, where each loudspeaker covers the whole audible frequency range (see also Section 4.4). Figures 4.17 and 4.18 show the condition number for a three-loudspeaker system. The result in Fig. 4.17 was obtained by placing two loudspeakers at angles $\alpha = \pm 5°$ and changing the azimuth angle of the third loudspeaker from 0° to 158° in 2° steps on a circle of radius 100 cm. Fig. 4.18 was obtained using the same procedure while the two fixed loudspeakers were positioned at angles $\alpha = \pm 45°$.

When the two fixed loudspeakers are positioned very close to each other, the matrix $\mathbf{C}(\omega)$ is ill-conditioned at low frequencies. Placing a third loudspeaker at $\alpha = 90°$ or at $\alpha = 270°$ improves the condition at low frequencies on the cost of some aliasing at higher frequencies. When the two fixed loudspeakers are positioned far from each other, the ma-

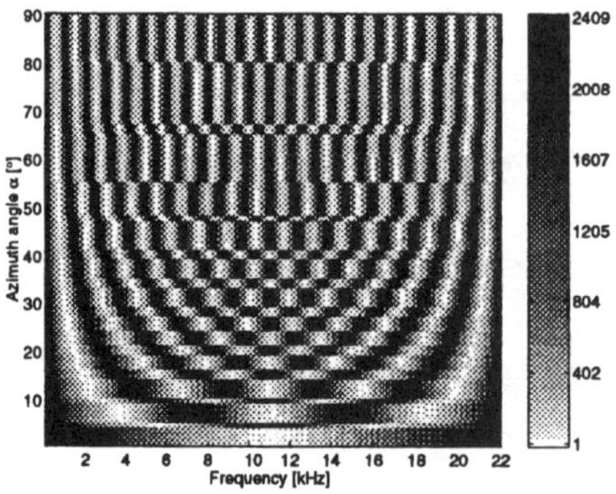

Figure 4.19: The condition number of $\mathbf{C}(\omega)$ in a room with $\{\beta_i = 0.2 : i = 1, 2, \cdots, 6\}$. d_m = 18 cm, and r = 100 cm.

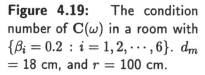

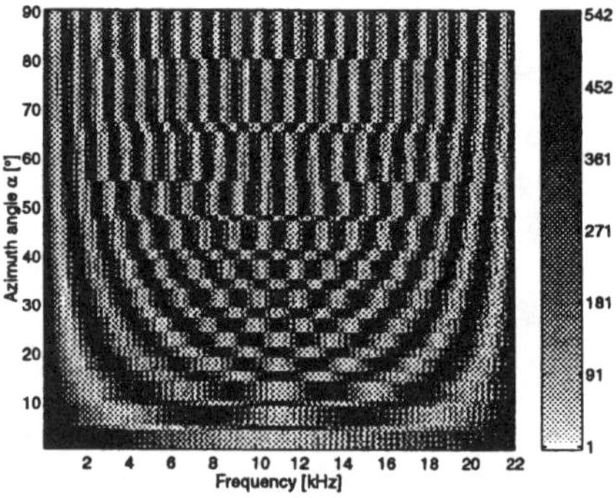

Figure 4.20: The condition number of $\mathbf{C}(\omega)$ in a room with $\{\beta_i = 0.5 : i = 1, 2, \cdots, 6\}$. d_m = 18 cm, and r = 100 cm.

trix $\mathbf{C}(\omega)$ is well conditioned only in small frequency bands between the peaks of the condition number. Placing a third loudspeaker at $\alpha = 0°$ or at $\alpha = 180°$ removes most of the peaks and, therefore, widens the frequency band in which the matrix is well-conditioned. This optimum three-loudspeaker set-up has been used in the experiment presented in Fig. 4.12.

4.5.3 Performance in Reverberant Environments

In most studies of sound reproduction systems, only simplified free field acoustic transfer functions are used. Although this leads to better understanding of the underlying physical principles, sound reproduction systems are rarely designed to operate in free field environments. Therefore, examination of the effects of reverberation on results obtained under free field conditions is necessary. This is readily performed in the experimental set-up used in the present work by changing the reflection coefficients of the room boundaries in the room simulation program. The results of these changes are shown in Fig. 4.19 and Fig. 4.20 when all β_i are set to 0.2 and 0.5, respectively.

At moderate reverberation, the traces of the condition number with respect to the frequency are similar to those in the anechoic case. However, the maximum value of the condition number increases significantly. At some frequencies the condition number is nearly 2500. The condition

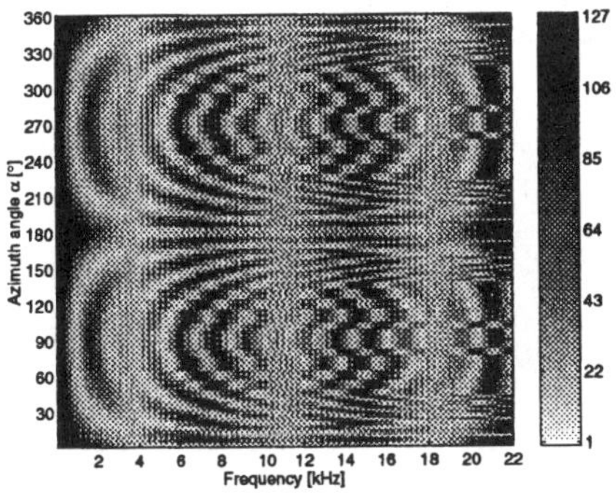

Figure 4.21: The condition number of $\mathbf{C}(\omega)$ in a room with $\{\beta_i = 0.5 : i = 1, 2, \cdots, 6\}$ for three loudspeakers. Two loudspeakers are placed at azimuth angles $\alpha = \pm 5°$ while the position of the third is variable. d_m = 18 cm, and r = 100 cm.

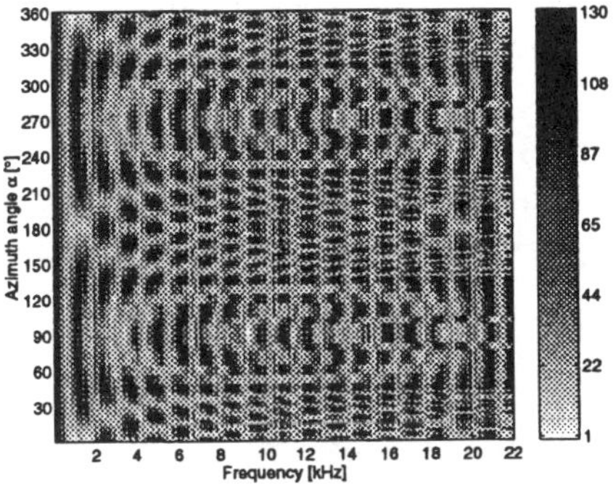

Figure 4.22: The condition number of $\mathbf{C}(\omega)$ in a room with $\{\beta_i = 0.5 : i = 1, 2, \cdots, 6\}$ for three loudspeakers. Two loudspeakers are placed at azimuth angles $\alpha = \pm 45°$ while the position of the third is variable. d_m = 18 cm, and r = 100 cm.

number does not decrease with increasing spatial aliasing at larger α as in the anechoic case. In more reverberant environments, not only the increase in condition number is noticed, but the traces are also smeared over the neighbouring frequency bands. The peaks of the condition number are also higher and wider. This may suggest that increasing reverberation in the listening environment leads to decreasing the system robustness to listener movements.

The situation is different in the case of three-loudspeaker systems. This can be seen from Fig. 4.21 and Fig. 4.22 which are obtained by repeating the experiments shown in Fig. 4.17 and Fig. 4.18 after setting all β_i to 0.5. In the case of two closely separated loudspeakers, little change is noticed by increasing reverberation. When the two fixed loudspeakers are widely separated, the condition number increases while the third loudspeaker has limited the smearing of traces over frequency. In general, a three-loudspeaker system is more robust in reverberant environments than a two-loudspeaker system.

4.6 Large Movements and Head Tracking

Up to this point, only small listener's movements have been considered. For small movements, it is sufficient to design the control filters to have average solutions that are valid within small areas around the listener's ears. When the listener moves beyond this area, the control filters become in great error, and new filters that are derived from the new listener's position must be used. Conventional binaural synthesis systems solve the problem of a moving listener (or a moving virtual source) by switching between pre-measured HRTF pairs [25]. To simulate a moving source, the system switches between stored HRTF pairs that are previously measured at discrete positions on the required source contour in the correct sequence. For a moving listener, a sensor is employed to continuously monitor the listener's position and send the position information to a computer. At every time moment, the position information is used to address the HRTF pair nearest to the current listener's position. When the HRTF set is sparsely sampled in space, interpolation is often used to calculate the HRTF pair at intermediate positions using the nearest available two or more HRTF pairs to the current position. In loudspeaker displays, both the cross-talk cancellation and the binaural synthesis filters must be switched according to the current listener's position.

Although acoustical, mechanical, and optical sensors are used in commercially available head trackers, most binaural synthesis systems make use of magnetic sensors, partially because they allow the listener to move 360° and partially for their low cost. Recently, more interest is directed to video-based head tracking systems [72, 95]. In such systems, a video camera monitors the listener's position, which frees the listener from wearing any hardware. When a video camera is used, image processing can be employed to further improve the system accuracy. One approach, presented in [95], is to store several complete sets of HRTFs measured for human listeners, together with an image of each listener's pinna. The system then chooses the most suitable set of HRTFs for the current user based on a comparison between the image of the user's pinna and the stored pinna images.

In an adaptive audio reproduction system, the control filters are designed in real-time, and the above mentioned switching technique can not be used. Provided that the adaptive filters can be updated faster than the listener's movements, it is sufficient to readapt the control filters to their new optimum solutions that correspond to the current listener's position. This is complicated by the following factors,

- The filtered-x LMS and the adjoint LMS algorithms used to update the control filters (see Chapter 3) require estimating all the electro-acoustic transfer functions $\underline{\mathbf{C}}_{ml}(\omega)$.
- Adaptive 3D sound systems introduced in Chapter 3 are based on cancelling the sound emitted from a physical loudspeaker at the listener's ears. Using the same technique with the audio signals results in cancelling the sound meant for the listener to hear.
- The filtered-x and adjoint LMS algorithms are known for their slow conversion speed due to the electro-acoustic transfer functions in front of the filters.

These issues are the subjects of the next few sections. On-line estimation of electro-acoustic transfer functions is addressed in Section 4.7. Methods for updating the control filters while leaving the audio signal for the listener to hear are discussed in Section 4.7.3. Finally, improving the convergence speed of the filtered-x and the adjoint LMS algorithms is treated in Section 4.8.

4.7 On-Line Identification

Both the MEFX and the ALMS algorithms discussed in Chapter 3 require measuring the acoustic transfer functions between the reproduction loudspeakers and the microphones. In a static situation, measuring these transfer functions off-line in an initialisation stage is sufficient. In a dynamic situation when one (or more) of these transfer functions changes, new measurements must be performed and used to recalculate the matrix of control filters $\mathbf{W}(\omega)$. When such changes occur frequently, it is necessary to continuously measure the electro-acoustic transfer functions and use the last measurement to update the filters. Methods to measure the transfer functions while updating the control filters are discussed in

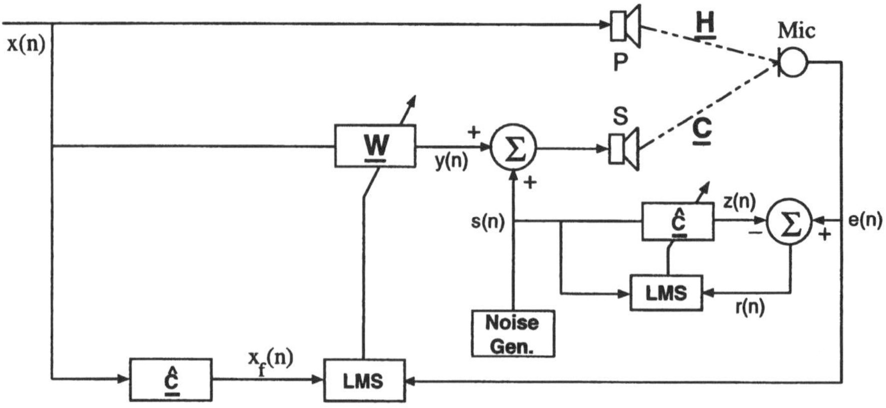

Figure 4.23: The filtered-x LMS algorithm with on-line identification using an extra reference signal.

this section. Two main methods are known in active noise control literature for on-line identification. The first uses an extra reference signal to update an adaptive filter connected in parallel with the acoustic transfer function to be measured. This method is discussed in Section 4.7.1. The second uses no extra reference signal and is discussed in Section 4.7.2. The application of these methods to multichannel audio reproduction systems is investigated in Section 4.7.3.

4.7.1 Using Extra Reference Signals

In both off-line and on-line identification of acoustic transfer functions, an adaptive identification process is usually employed. Although adaptive filters connected in series with the electro-acoustic transfer function to be measured may be used [61], only FIR adaptive filters connected in parallel are considered here. The basic principle of on-line identification using extra reference signals is introduced in Section 4.7.1.1 for a single channel system. Improving the convergence speed and extending the basic principle to the multichannel case are discussed in Section 4.7.1.2 and Section 4.7.1.3, respectively.

4.7.1.1 The Basic Principle

The basic principle of on-line adaptive identification using an extra white noise reference signal [60, 61] is shown in Fig. 4.23 for a single channel active noise control system. The filtered-x (or alternatively the adjoint) LMS algorithm is used to update the control filter $\underline{\mathbf{W}}$ to cancel the sound emitted by loudspeaker P at the microphone position. At the same time, another adaptive filter $\underline{\hat{\mathbf{C}}}$ is used to estimate the acoustic transfer function $\underline{\mathbf{C}}$. An extra white noise signal $s(n)$ (or any spectrally rich alternative such as a chirp signal) that is uncorrelated with $x(n)$ is added to the control signal $y(n)$ and the sum is used to drive loudspeaker S. The superposition of the responses of loudspeakers P and S is observed by the microphone Mic. The microphone signal $e(n)$ can be expressed as

$$e(n) = s(n) * \underline{\mathbf{c}} + x(n) * [\underline{\mathbf{h}} + \underline{\mathbf{w}} * \underline{\mathbf{c}}], \tag{4.39}$$

which shows that $e(n)$ does not only contain the desired component $s(n) * \underline{\mathbf{c}}$ but is also contaminated with a disturbance term which is dependent on $x(n)$. The output of the filter $\underline{\hat{\mathbf{C}}}$ is subtracted from the microphone output and the result $r(n) = e(n) - z(n)$ is used as the error, the square of which is to be minimised by the LMS algorithm. The input to the adaptive filter $\underline{\hat{\mathbf{C}}}$ and its LMS update algorithm is the extra white noise reference signal $s(n)$ only. Since $s(n)$ is chosen to be uncorrelated with $x(n)$, the adaptive filter $\underline{\hat{\mathbf{C}}}$ converges on average to the desired response $\underline{\mathbf{C}}$, regardless of the above mentioned disturbance [144]. However, the disturbance degrades the convergence speed of the filter, a matter that will be shortly discussed in Section 4.7.1.2.

The main disadvantage of the above mentioned method is that the extra reference signal remains to be heard at the microphone. In active noise cancellation applications, this represents an additional residual noise, which may be tolerated if the reference signal $s(n)$ is about 25 dB lower in level than $x(n)$.

4.7.1.2 Improving the Convergence Speed

The reference signal $s(n)$ in Fig. 4.23 must be kept at a low level for adequate system performance. The identification error $r(n)$ is, therefore, contaminated with a large disturbance due to the components $d(n) =$

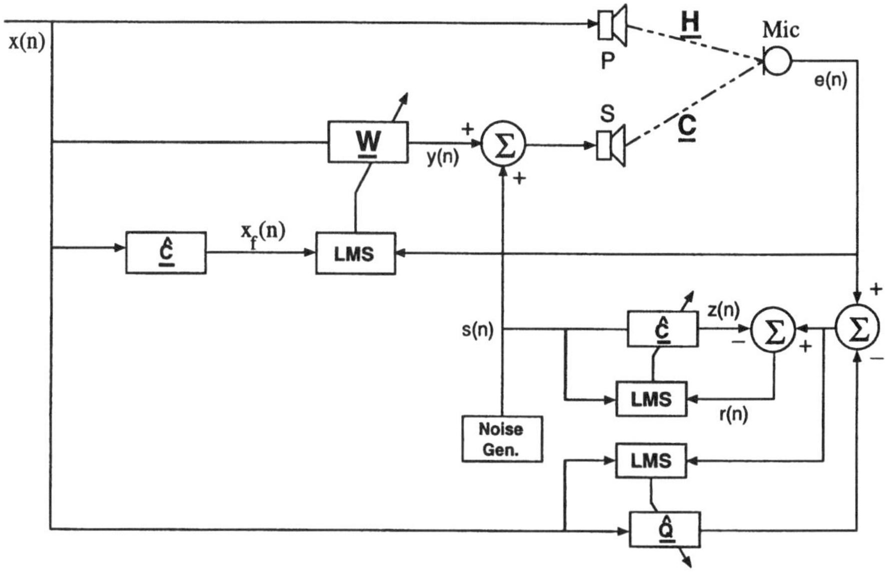

Figure 4.24: Improving the convergence of the on-line identification process.

$x(n) * \underline{\mathbf{h}}$ and $\hat{d}(n) = y(n) * \underline{\mathbf{c}}$. For a stable identification process, the step size μ of the LMS update of $\hat{\underline{\mathbf{C}}}$ must be kept small [93, 144], which results in a slow convergence speed. The convergence speed can be improved by estimating the undesired component $d(n) + \hat{d}(n)$ and subtracting it from $e(n)$. This is achieved using a third adaptive filter $\hat{\underline{\mathbf{Q}}}$ with input $x(n)$ to estimate the transfer function $\underline{\mathbf{h}} + \underline{\mathbf{w}} * \underline{\mathbf{c}}$ as shown in Fig. 4.24. In this case, $s(n) * \underline{\mathbf{c}}$ represents a disturbance for the adaptive filter $\hat{\underline{\mathbf{Q}}}$. Since $s(n)$ and $x(n)$ are uncorrelated, and the variance of $s(n)$ is much smaller than that of $x(n)$, $\hat{\underline{\mathbf{Q}}}$ converges fast to its average final solution. This in turn cancels the disturbance component in $r(n)$, allowing the use of a larger step size to update $\hat{\underline{\mathbf{C}}}$.

4.7.1.3 Multichannel Identification

In multichannel systems, all transfer functions between the reproduction loudspeakers and the microphones $\{\underline{\mathbf{C}}_{ml} : m = 1, 2, \cdots, M, l = 1, 2, \cdots, L\}$ must be estimated. However, it is not possible to estimate all $M \times L$ channels simultaneously using one extra reference signal $s(n)$

due to the interchannel coupling effects [93]. Due to this coupling, the estimate of $\underline{\mathbf{C}}_{ml}$ is biased by the estimation error of other transfer functions from all loudspeakers to the same microphone m. For this reason, only transfer functions from one of the loudspeakers to all microphones $\{\underline{\mathbf{C}}_{ml} : m = 1, 2, \cdots, M\}$ may be identified simultaneously using a single extra reference signal. This identification process is then repeated for the other loudspeakers, which increases the identification time. One approach to speed up the identification of multichannel systems is to employ L independent reference signals, one for each reproduction loudspeaker. Alternatively, a single reference signal may be used with inter-channel delay to decorrelate the excitation signals to the loudspeakers [94]. The latter method is more efficient, since only one internal reference signal must be generated compared to L signals in the former.

4.7.2 Using the Control Signals

Using an extra reference signal for on-line identification as described in Section 4.7.1 may not be desired in some applications. In such cases, the control signal $y(n)$ may be used as a reference signal. In general, the reference signal $y(n)$ is not persistent and its spectrum is not flat. The spectral contents of such a reference signal depends on the spectral contents of the input signal $x(n)$ and the frequency response of the control filter $\underline{\mathbf{W}}$. This leads to a slow convergence speed, and the estimated filter $\hat{\underline{\mathbf{C}}}$ depends on the spectrum of $y(n)$. However, this approach may be attractive in many situations, especially when the adaptation is performed in the frequency domain, allowing power normalisation in each frequency bin as mentioned in Section 3.4.2.

Another problem that arises when using $y(n)$ as a reference signal is that $y(n)$ is always well correlated with $x(n)$. The microphone signal in this case is given by

$$e(n) = \underline{\mathbf{x}}^T(n)\ \underline{\mathbf{h}} + \underline{\mathbf{y}}^T(n)\ \underline{\mathbf{c}}, \tag{4.40}$$

which shows that the microphone signal is contaminated with a component that is dependent on $x(n)$. For proper convergence, the disturbance $\underline{\mathbf{x}}^T(n)\ \underline{\mathbf{h}}$ in the microphone signal must be removed. This may be achieved by using an additional adaptive filter similar to the approach followed to improve the convergence speed in Section 4.7.1.2.

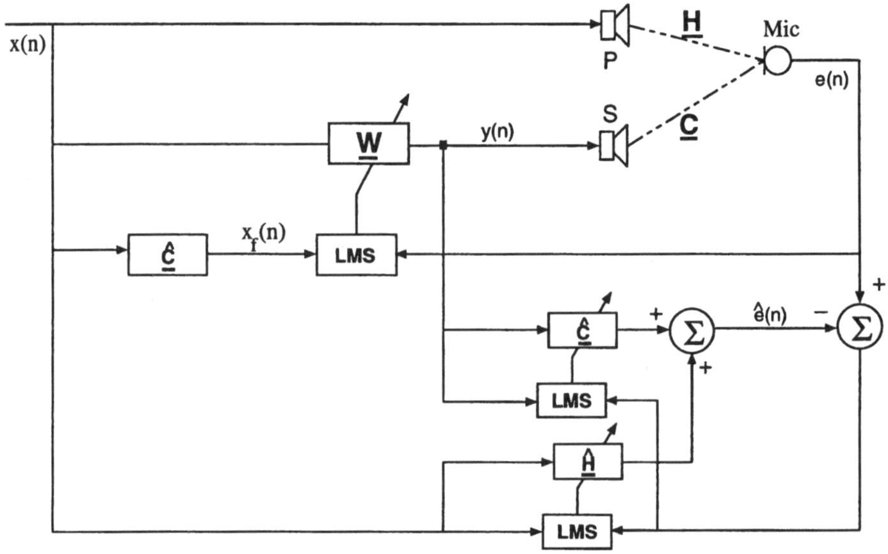

Figure 4.25: The filtered-x LMS algorithm with on-line identification using the control signal $y(n)$ as a reference signal.

The block diagram of a single channel active noise control system employing the control signal as a reference signal is shown in Fig. 4.25. The acoustic transfer function $\underline{\mathbf{C}}$ is estimated using the adaptive filter $\hat{\underline{\mathbf{C}}}$. Another adaptive filter $\hat{\underline{\mathbf{H}}}$ is used to estimate the transfer function between loudspeaker P and the microphone. The outputs of the two estimation filters are added to obtain an estimate $\hat{e}(n)$ of the microphone signal $e(n)$. The difference between $e(n)$ and $\hat{e}(n)$ is used to update both $\hat{\underline{\mathbf{C}}}$ and $\hat{\underline{\mathbf{H}}}$. In this structure, the two estimation filters $\hat{\underline{\mathbf{C}}}$ and $\hat{\underline{\mathbf{H}}}$ are considered as one long filter composed of two parts. The first part $\hat{\underline{\mathbf{H}}}$ has $x(n)$ as its reference signal while the second part $\hat{\underline{\mathbf{C}}}$ has $y(n)$ as a reference signal. This can be seen by rewriting (4.40) as a vector multiplication as follows

$$e(n) = \left[\begin{array}{cc} \underline{\mathbf{x}}^T(n) & \underline{\mathbf{y}}^T(n) \end{array} \right] \left[\begin{array}{c} \underline{\mathbf{h}} \\ \underline{\mathbf{c}} \end{array} \right]. \qquad (4.41)$$

The condition for successful identification may be obtained by left multiplying (4.41) once by $\underline{\mathbf{x}}(n)$ and once by $\underline{\mathbf{y}}(n)$ and taking the expectation

[125]

$$\mathrm{E}\begin{bmatrix} \underline{\mathbf{x}}(n)\, e(n) \\ \underline{\mathbf{y}}(n)\, e(n) \end{bmatrix} = \mathrm{E}\begin{bmatrix} \underline{\mathbf{x}}(n)\, \underline{\mathbf{x}}^T(n) & \underline{\mathbf{x}}(n)\, \underline{\mathbf{y}}^T(n) \\ \underline{\mathbf{y}}(n)\, \underline{\mathbf{x}}^T(n) & \underline{\mathbf{y}}(n)\, \underline{\mathbf{y}}^T(n) \end{bmatrix} \begin{bmatrix} \underline{\mathbf{h}} \\ \underline{\mathbf{c}} \end{bmatrix}. \tag{4.42}$$

The unknown filters $\underline{\mathbf{h}}$ and $\underline{\mathbf{c}}$ may, therefore, be calculated only if the correlation matrix given by the first factor of the right hand side of (4.42) is non-singular. This matrix is non-singular if its rank is $N_h + N_c$, where N_h and N_c are the length of $\underline{\mathbf{h}}$ and $\underline{\mathbf{c}}$, respectively. This is fulfilled only if the length of the control filter $\underline{\mathbf{W}}$ satisfies $N_w > N_h - N_c + 1$ [125]. This approach provides continuous tracking of both $\underline{\mathbf{H}}$ and $\underline{\mathbf{C}}$ simultaneously, which is advantageous for a frequently moving microphone since both transfer functions change frequently. On the other hand, if the frequent changes are encountered only in $\underline{\mathbf{C}}$ while $\underline{\mathbf{H}}$ remains constant, the estimation of $\underline{\mathbf{H}}$ represents extra overhead calculations.

When using the control signals as reference signals for on-line identification, simultaneous identification of multichannel transfer functions such as that mentioned in Section 4.7.1.3 is not possible. This is due to the fact that all reference signals $\{y_{ml}(n) : l = 1, 2, \cdots, L\}$ are correlated with $x(n)$ and, therefore, correlated with each other. Only simultaneous identification of the transfer functions $\{\underline{\mathbf{c}}_{ml} : m = 1, 2, \cdots, M\}$ is possible. This is also true for multiple input signals, since each of the L control signals $\{y_{ml}(n) : l = 1, 2, \cdots, L\}$ are correlated with all input signals $\{x_k(n) : k = 1, 2, \cdots, K\}$.

4.7.3 Audio Reproduction and On-Line Identification

In a sound reproduction system and for large listener's movements, the control filters are in great error and must be redesigned as mentioned before. Adaptive sound reproduction systems may redesign the filters by readapting them to their new optimum solutions at the new listener's position. In this section, on-line adaptation of the control filters while the audio reproduction system is in operation is addressed. The new solutions require knowledge of the new electro-acoustic transfer functions $\mathbf{C}$ between the reproduction loudspeakers and the current positions of the listener's ears. When microphones are placed in the listener's ear canals, the above mentioned on-line identification techniques may be used to continuously measure the electro-acoustic transfer functions $\mathbf{C}$.

In some applications, such as cross-talk cancellation, knowledge of the current $\mathbf{C}$ alone is sufficient to readapt the filters as discussed in Section 4.7.3.1. In other applications, such as virtual sound source synthesis, knowledge of the current matrix $\mathbf{H}$ is also required as will be shortly shown in Section 4.7.3.2.

4.7.3.1 Tracking for Cross-Talk Cancellation

The desired response $d(n)$ in the single channel cross-talk cancellation system shown in Fig. 4.26 is a delayed version of the reference signal $x(n)$ as mentioned in Section 3.3.7. The system response at the microphone $\hat{d}(n)$ is the sound signal that the listener actually hears. The error $e(n)$ is formed by *electrically* subtracting the microphones' output from the desired response. This is in contrast with other applications such as active noise control where both $d(n)$ and $\hat{d}(n)$ are acoustic waves that are added by the microphone. This simplifies both the on-line identification of $\underline{\mathbf{C}}$ and the on-line adaptation of $\underline{\mathbf{W}}$ for listener tracking. Furthermore, when the listener moves, only $\underline{\mathbf{C}}$ has to be re-estimated, since $\underline{\mathbf{H}}$ is always a known pure delay.

Consider first the on-line identification of the electro-acoustic transfer function $\underline{\mathbf{C}}$. Since the microphone's output is not contaminated with any primary disturbances, the identification process is considerably simplified. No estimation of such disturbances need to be performed as in the general case described above. This simplification is shown in Fig. 4.26 for a single channel case when $y(n)$ is used as a reference signal for the identification process. Using an extra reference signal for identification results in a slower convergence speed due to the disturbance component from $y(n)$ at the microphone as mentioned in Section 4.7.1. This component is also readily estimated and subtracted from the microphone output since only the transfer function $\underline{\mathbf{C}}$ is involved.

Turning the attention to the on-line adaptation of $\underline{\mathbf{W}}$, since $d(n)$ is in electric rather than acoustic form, subtracting $\hat{d}(n)$ from $d(n)$ does not null the acoustical waves that are meant for the listener to hear. Therefore, the audio signal $x(n)$ may be used as a reference signal for the adaptation of the filter $\underline{\mathbf{W}}$ as shown in Fig. 4.26. This may not be an option in other applications. A work around is discussed in Section 4.7.3.2 for virtual source synthesis systems.

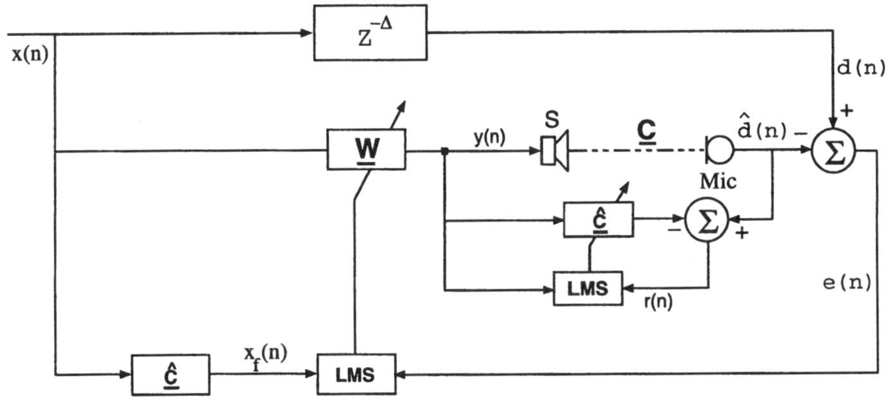

Figure 4.26: Cross-talk cancellation with an integrated on-line identification for listener tracking.

4.7.3.2 Tracking for Virtual Source Synthesis

A single channel virtual source synthesis system is shown in Fig. 4.27. The adaptive filter $\underline{\mathbf{W}}$ in this system is designed off-line by playing a white noise signal $s(n)$ through loudspeaker P and adjusting the filter coefficients to cancel the resulting sound at the microphone as mentioned in Section 3.3.6. Clearly, if this approach is used for on-line adaptation of $\underline{\mathbf{W}}$ with $s(n)$ as the music signal required to be played by the virtual source, the adaptation process will cancel the music at the microphone, and the listener will hear silence.

The straightforward approach to on-line adaptation of $\underline{\mathbf{W}}$ is to use an extra reference signal similar to the on-line identification mentioned in Section 4.7.1. The music signal $x(n)$ is multiplied by -1 to compensate for the negative sign in the optimum solution and added at the input of the filter $\underline{\mathbf{W}}$. When the filter adapts to its optimal solution, the acoustic response due to $x(n)$ at the microphone equals $\underline{\mathbf{h}} * x(n)$, which is the desired response for creating a virtual sound source at the position of loudspeaker P. The control signal $y(n)$ is used as a reference signal for the on-line identification of $\underline{\mathbf{C}}$ and $\underline{\mathbf{H}}$ as discussed in Section 4.7.2. The white noise signal $s(n)$ is the extra reference signal used for the on-line adaptation of the filter $\underline{\mathbf{W}}$. As not to degrade the quality of the music heard at the microphone, the noise $s(n)$ must be kept at a low amplitude. The music response at the microphone, therefore, forms a high

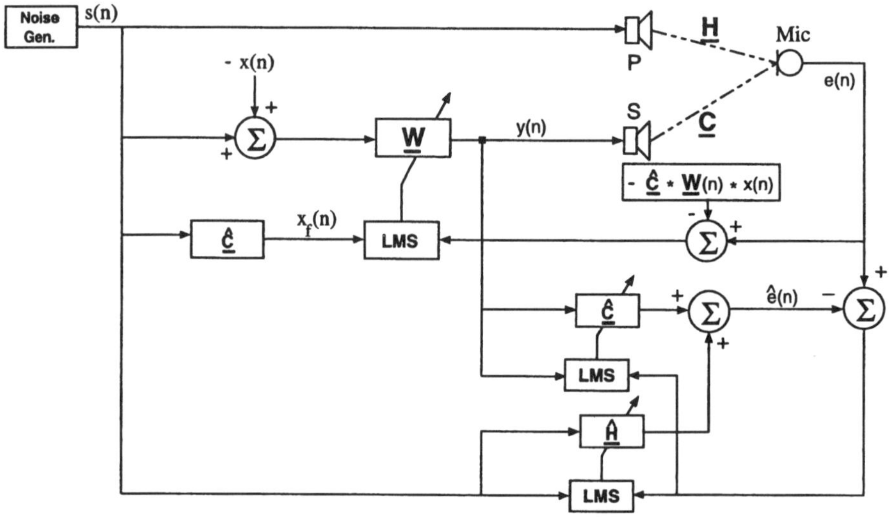

Figure 4.27: One channel of a virtual sound source synthesis system with on-line identification and on-line adaptation of the control filter.

level disturbance to the adaptation process of $\underline{\mathbf{W}}$. This can be alleviated by subtracting an estimate of this disturbing component from the microphone signal before using it in updating $\underline{\mathbf{W}}$ as shown in Fig. 4.27. Unlike in active noise control applications, the acoustic response due to the extra reference noise at the microphone is masked by the much louder music. Therefore, the upper limit of the amplitude of $s(n)$ is defined by the masking properties of the human auditory system [148].

The above mentioned system assumes that a physical loudspeaker P exists at the position where a virtual sound source is required. In a real-time virtual sound source synthesis system, such a physical loudspeaker is absent and the architecture shown in Fig. 4.27 can not be used. However, this system (with $x(n)$ absent) is very efficient for calculating and storing a set of filters $\underline{\mathbf{W}}$ for later use in virtual source simulation at different positions. A head tracking system may then be used to select one of the pre-calculated set of filters, through which the music signal should be filtered in real-time.

In many virtual sound source synthesis systems, the matrix of electro-acoustic transfer functions $\mathbf{H}$ is available as a pre-measured and stored set of HRTFs from different directions. In such systems, the desired responses $\underline{\mathbf{d}}(n)$ may be calculated internally by filtering the music signals $\underline{\mathbf{x}}(n)$ through the appropriate set of transfer functions $\mathbf{H}$, and the error vector $\underline{\mathbf{e}}(n)$ is calculated *electrically* as in the case of cross-talk cancellation. This allows the use of music signals as reference signals for the adaptive filters $\mathbf{W}$ similar to the structure shown in Fig. 4.26. Unlike in cross-talk cancellation, the optimum solution $\underline{\mathbf{W}}_{opt}$ in this case is dependent on both $\mathbf{H}$ and $\mathbf{C}$. Provided that a switching mechanism is integrated in the system to select the appropriate new matrix $\mathbf{H}$ for every listener movement, the on-line identification may be employed to measure $\mathbf{C}$ and the filters $\underline{\mathbf{W}}$ are adapted to the optimum solution for the new listener position. The need for a head tracking system may be avoided if the music signals are stored after processing them by the HRTFs. This reduces the system to a cross-talk canceller and the system shown in Fig. 4.26 may be used.

4.8 Fast Adaptive Algorithms

The convergence speed of the single channel filtered-x LMS algorithm discussed in Section 3.3.1 is known to be slow due to the electro-acoustic transfer function between the secondary loudspeaker and the microphone (the secondary acoustic path) in front of the adaptive filter. The same also holds for the adjoint LMS algorithm discussed in Section 3.4.1. It is shown in [144] that the convergence properties of the LMS adaptive algorithm can be improved by decorrelating its input signal. This process results in convergence properties that are independent of the input signal statistics. However, the decorrelation process is not obvious in the filtered-x LMS (FX) case, since the update suffers from coloration not only due to the input signal but also due to the estimate of the secondary acoustic path used to calculate the filtered-x signal. In this section, two methods are presented to improve the convergence properties of that algorithm [67, 132]. A straightforward application of the decorrelation principle to the update equation of the FX algorithm leads to the decorrelated filtered-x LMS (DFX) algorithm described in Section 4.8.1. The DFX algorithm suffers from the problem of division by a small number, which is enhanced by the amplitude spectrum of the secondary acoustic path. Partially decorrelating the filtered input signal

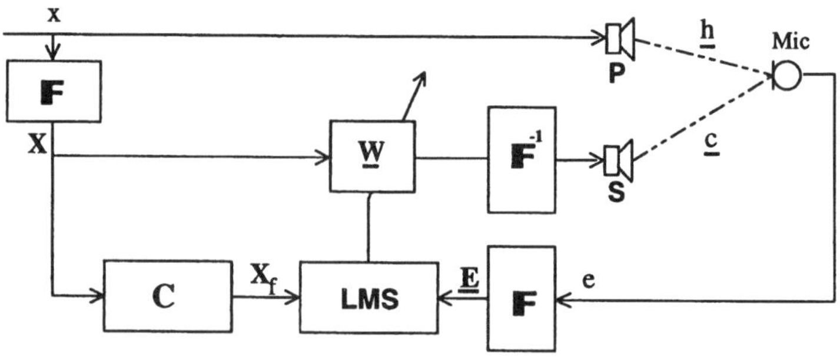

Figure 4.28: A single channel noise canceller in the frequency domain.

leads to the all-pass filtered-x (APFX) algorithm presented in Section 4.8.2. The convergence properties of the different decorrelation methods are compared in Section 4.8.3.

4.8.1 The Decorrelated Filtered-X Algorithm (DFX)

The DFX algorithm is derived by decorrelating the conventional FX algorithm shown in Fig. 4.28. Since the decorrelation process is best explained in the frequency domain, the signals in Fig. 4.28 are transformed to the frequency domain, using the $\mathbb{F}$ blocks, where filtering and adaptation operations are performed. The update block in this figure uses a filtered version of the input signal $\mathbf{X}_f$ to adjust the vector of adaptive filter weights $\underline{\mathbf{W}}$ such that, on average, the square of the residual signal $\underline{\mathbf{E}}$ measured at the microphone position goes to a minimum. The filtered-x signal $\mathbf{X}_f$ is calculated by filtering the input signal $\mathbf{X}$ by an estimate of the secondary acoustic path $\mathbf{C} = \text{diag}\{\mathbb{F}\ \underline{\mathbf{c}}\}$, which is considered here to be a perfect estimate,

$$\mathbf{X}_f = \mathbf{X}\,\mathbf{C}. \tag{4.43}$$

With $\underline{\mathbf{H}} = \mathbb{F}\ \underline{\mathbf{h}}$, the residual signal $\underline{\mathbf{E}}$ can be calculated from the figure to be

$$\underline{\mathbf{E}} = \mathbf{X}\,(\mathbf{C}\ \underline{\mathbf{W}} + \underline{\mathbf{H}}), \tag{4.44}$$

and the update equation for the conventional FX algorithm is given by

$$\underline{\mathbf{W}}_{new} = \underline{\mathbf{W}}_{old} - 2\mu\ \mathbf{X}_f^*\ \underline{\mathbf{E}}, \tag{4.45}$$

where $\cdot^*$ is the complex conjugate operator and μ is the adaptation constant. Substituting (4.43) and (4.44) in (4.45) and using the assumption that all transfer functions are time invariant (or slowly time varying) gives

$$\underline{\mathbf{W}}_{new} = \underline{\mathbf{W}}_{old} - 2\mu\ \mathbf{C}^*\,\mathbf{X}^*\,\mathbf{X}\,(\mathbf{C}\ \underline{\mathbf{W}}_{old} + \underline{\mathbf{H}}). \tag{4.46}$$

Applying the expectation operator on both sides of (4.46) and denoting $E\{\mathbf{C}^*\,\mathbf{X}^*\,\mathbf{X}\,\mathbf{C}\}$ by $\mathbf{P}_{X_f} = \text{diag}\{\underline{\mathbf{P}}_{X_f}\}$, the diagonal power matrix of $\mathbf{X}_f$, and rearranging gives the dynamic convergence equation for the conventional FX

$$\underline{\widetilde{\mathbf{W}}}_{new} = (\mathbf{I} - 2\mu\ \mathbf{P}_{X_f})\ \underline{\widetilde{\mathbf{W}}}_{old} - 2\mu\ \mathbf{P}_{X_f}\ (\mathbf{C}^{-1}\ \underline{\mathbf{H}}), \tag{4.47}$$

in which $\underline{\widetilde{\mathbf{W}}}$ denotes $E\{\underline{\mathbf{W}}\}$ and $\mathbf{I}$ is the identity matrix. Equation (4.47) shows that after successful conversion, $\underline{\widetilde{\mathbf{W}}}_{new} \approx \underline{\widetilde{\mathbf{W}}}_{old}$ and the adaptive filter approaches its final solution $\underline{\widetilde{\mathbf{W}}}_{fin} \approx -\mathbf{C}^{-1}\ \underline{\mathbf{H}}$, which is the required solution for the noise canceller. Equation (4.47) also shows that the trajectory taken by $\underline{\widetilde{\mathbf{W}}}$ from its initial value to its final solution is governed by the term $(\mathbf{I} - 2\mu\ \mathbf{P}_{X_f})$. The slope taken by each filter coefficient is dependent on the power contained in $\mathbf{X}_f$ at the corresponding frequency bin. When the ratio $\max\{\underline{\mathbf{P}}_{X_f}\}/\min\{\underline{\mathbf{P}}_{X_f}\}$ is high, some filter coefficients converge to their final values much slower than the others. This is known as the eigenvalue disparity problem [144]. This problem is enhanced in the FX case, since the spectrum of $\mathbf{X}_f$ is colored not only by the input signal $\mathbf{X}$ but also by the estimate of the secondary acoustic path $\mathbf{C}$. This latter includes, apart from the acoustic response between the secondary source and the microphone, the response of the secondary loudspeaker, the microphone, the ADC and the DAC used in the identification process. The former two have band-pass characteristics and the latter two contain anti-aliasing and reconstruction filters, which make $\mathbf{C}$ a band-pass process with high energy ratio between mid-band and out-of-band frequencies. It is those frequencies with low energy that degrade the convergence properties of the conventional FX algorithm.

To improve the convergence, this coloration must be removed, which is usually done by multiplying the second term in (4.45) by the inverse of a decorrelation matrix $\mathbf{P}$, which leads to the update equation of the DFX

$$\underline{\mathbf{W}}_{new} = \underline{\mathbf{W}}_{old} - 2\mu\ \mathbf{P}^{-1}\ \mathbf{X}_f^*\,\underline{\mathbf{E}}. \tag{4.48}$$

Following the same steps as above, the dynamic convergence equation for the DFX can be obtained

$$\underline{\widetilde{\mathbf{W}}}_{new} = (\mathbf{I} - 2\mu\, \mathbf{P}^{-1}\, \mathbf{P}_{X_f})\, \underline{\widetilde{\mathbf{W}}}_{old} - 2\mu\, \mathbf{P}^{-1}\, \mathbf{P}_{X_f}\, (\mathbf{C}^{-1}\, \underline{\underline{\mathbf{H}}}). \qquad (4.49)$$

It follows from (4.49) that for all filter coefficients to approach their final values with the same speed, the decorrelation matrix should be chosen such that $\mathbf{P} = \mathbf{P}_{X_f}$ and (4.49) becomes

$$\underline{\widetilde{\mathbf{W}}}_{new} = (\mathbf{I} - 2\mu)\, \underline{\widetilde{\mathbf{W}}}_{old} - 2\mu\, (\mathbf{C}^{-1}\, \underline{\underline{\mathbf{H}}}), \qquad (4.50)$$

which shows a convergence behaviour that is dependent neither on the input signal $\mathbf{X}$ nor on the secondary acoustic path $\mathbf{C}$.

Although the DFX algorithm as given by (4.48) and (4.50) is completely decorrelated, it requires calculating and multiplying by the matrix $\mathbf{P}_{X_f}^{-1}$. In practice, this is done by maintaining the time average of a power vector $\underline{\mathbf{P}}_{X_f}$ and dividing the adaptation constant by that vector. In most practical cases, $\underline{\mathbf{P}}_{X_f}$ contains elements of small values due to the band-pass nature of $\mathbf{C}$ as mentioned above. Since the power estimate is never accurate, dividing by such small numbers may apply large changes to the update constants and drive the update unstable. To avoid this instability, the regularisation principle is often applied, a simple form of which is to divide by $(\underline{\mathbf{P}}_{X_f} + \delta\, \underline{\mathbf{1}})$, where δ is a small scalar and $\underline{\mathbf{1}}$ is a vector of ones. This, however, limits the decorrelation of the slowest modes (corresponding to the low power frequencies) and degrades the convergence as will be shown in Section 4.8.3.

4.8.2 The All-Pass Filtered-X Algorithm (APFX)

Recognising that the slowest modes are due to the band-pass nature of $\mathbf{C}$, it is, therefore, preferred to handle the decorrelation of $\mathbf{C}$ differently. Assuming that the input signal is well behaved, it can still be decorrelated by multiplying by its inverse power matrix $\mathbf{P}_X^{-1}$. Decorrelation of $\mathbf{C}$ is then performed partly by multiplying by $|\mathbf{C}^*|^{-1}$ instead of $|\mathbf{C}^*\,\mathbf{C}|^{-1}$, so that the decorrelation matrix $\mathbf{P}^{-1}$ in (4.48) now equals $\mathbf{P}_X^{-1}\,|\mathbf{C}^*|^{-1}$ and the approximate decorrelated dynamic convergence equation becomes

$$\underline{\widetilde{\mathbf{W}}}_{new} = (\mathbf{I} - 2\mu\, \mathbf{P}_X^{-1}\, |\mathbf{C}^*|^{-1}\, \mathbf{P}_{X_f})\, \underline{\widetilde{\mathbf{W}}}_{old} - 2\mu\, \mathbf{P}_X^{-1}\, |\mathbf{C}^*|^{-1}\, \mathbf{P}_{X_f}\, (\mathbf{C}^{-1}\, \underline{\underline{\mathbf{H}}}). \qquad (4.51)$$

Substituting $\mathbf{P}_{X_f} = \mathbf{C}^* \mathbf{X}^* \mathbf{X} \mathbf{C}$ in (4.51) and simplifying, we obtain the dynamic convergence equation of the APFX algorithm

$$\underline{\widetilde{\mathbf{W}}}_{new} = (\mathbf{I} - 2\mu\, \mathbf{C}_a^* \mathbf{C})\, \underline{\widetilde{\mathbf{W}}}_{old} - 2\mu\, \mathbf{C}_a^* \mathbf{C}\, (\mathbf{C}^{-1}\, \underline{\mathbf{H}}), \tag{4.52}$$

in which $\mathbf{C}_a^* = |\mathbf{C}^*|^{-1}\, \mathbf{C}^*$ is an all-pass filter having a phase response equals to that of $\mathbf{C}^*$. Comparing (4.52) and (4.50) shows that the APFX is still colored by the amplitude response of $\mathbf{C}$ ($\mathbf{C}_a^*$ has a flat amplitude response and, therefore, no coloration effect). Despite this residual coloration, it is shown in section 4.8.3 that the convergence properties of the APFX algorithm are significantly improved compared to the conventional FX algorithm. The update equation of the APFX algorithm can be derived from (4.51) to be

$$\underline{\mathbf{W}}_{new} = \underline{\mathbf{W}}_{old} - 2\mu\, \mathbf{P}_X^{-1}\, \mathbf{X}_{fa}^*\, \underline{\mathbf{E}}, \tag{4.53}$$

where $\mathbf{X}_{fa} = \mathbf{C}_a \mathbf{X}$ is a new filtered-x signal, obtained by filtering the input signal $\mathbf{X}$ by the all-pass filter $\mathbf{C}_a$ and hence the algorithm's name. Thus, in the APFX, the partial decorrelation of $\mathbf{C}$ is performed by calculating the filtered-x signal using $\mathbf{C}_a$ instead of $\mathbf{C}$ and no division by the power spectrum of $\mathbf{C}$ is needed.

In many applications, it is known a priori that the input signal $\mathbf{X}$ has a flat amplitude spectrum. In these cases, it is sufficient to decorrelate $\mathbf{C}$ only by updating the filter coefficients using $\mathbf{X}_{fa}$ without multiplying by $\mathbf{P}_X^{-1}$. This improves the convergence properties significantly without any extra computations, since $\mathbf{C}_a$ can be estimated in the initialisation stage instead of $\mathbf{C}$ while decorrelation of the input spectrum costs calculating and multiplying by $\mathbf{P}_X^{-1}$ during the update process. Another special case of the APFX algorithm is when the input signal $\mathbf{X}$ contains the dominant slow modes or comparable slow modes to those in $\mathbf{C}$. In this case, decorrelation of $\mathbf{X}$ using $\mathbf{P}_X^{-1}$ suffers from the same problems discussed in Section 4.8.1 and the same partial decorrelation principle using phase information only may be performed.

4.8.3 Performance Comparison

The convergence properties of the above presented algorithms have been extensively tested by means of computer simulations. The simulations implemented the single channel noise canceller shown in Fig. 4.28 using Block Frequency Domain Adaptive Filters (BFDAF) [55, 130]. The performance was studied using different acoustic transfer functions for $\underline{c}$ and $\underline{h}$. The results presented here were obtained using KEMAR Head-Related Impulse Responses (HRIR) in the horizontal plane (see Section 2.1.7). The impulse response $\underline{c}$ was the HRIR of the left ear from azimuth angle $\alpha = 5°$, while $\underline{h}$ was that from $\alpha = 35°$. Each impulse response was an FIR filter of 512 coefficients measured at 44.1 kHz sampling frequency. Fig. 4.29 shows the time, amplitude, and phase responses of the secondary acoustic path $\underline{c}$. In the following simulations, the adaptive filter length was 1024, the new input samples in each block was 1025, and the Fourier transformation length was 2048. Each simulation updated the adaptive filter $\underline{\mathbf{W}}$, starting from zero initial condition, according to one of the above presented algorithms, and the mean squared error (time average of $10 \log_{10}(\mathrm{r}^2)$) was plotted against time. To perform honest comparison, the adaptation constant μ was adjusted to give the same theoretical final misadjustment with all algorithms.

4.8.3.1 Performance for a White Noise Reference Signal

With a white noise reference signal $\mathbf{X}$ of zero mean and variance 1, the adaptive filter $\underline{\mathbf{W}}$ was updated according to each of the above presented algorithms. The learning curves for three specific cases are shown in Fig. 4.30. In this figure, curve 1 is the learning curve for the conventional FX algorithm given by (4.45). Curve 2 is that for the APFX given by (4.53). Since the reference signal is white noise, multiplication by $\mathbf{P}_X^{-1}$ does not count, and was not performed. Curve 3 is the learning curve for the DFX update given by (4.48) with $\mathbf{P}_{X_f} + \delta\,\underline{1}$ and $\delta = 0.001$.

From Fig. 4.30, the improvement achieved by the APFX over the conventional FX algorithm is evident. This is attributed to the considerable coloration added to the filter's input signal by $\underline{c}$ and partially removed by using the APFX algorithm. From experiments using other transfer functions [67], it is observed that the improvement obtained by the APFX over FX increases as the coloration introduced by $\underline{c}$ increases.

In this experiment, it was noticed that updating the filter according to the DFX without regularisation, or with δ smaller than about 10^{-5}, results in an unstable update. This may be attributed to the pinna notch

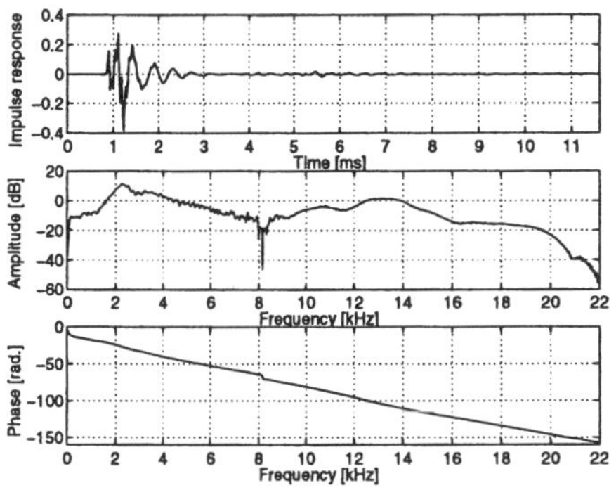

Figure 4.29: The secondary acoustic transfer function.

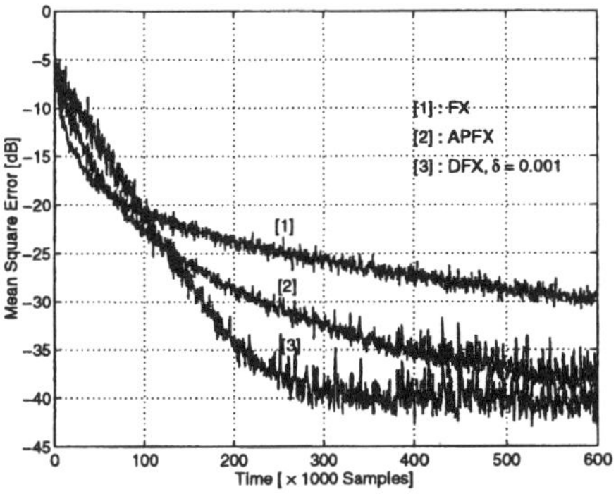

Figure 4.30: Learning curves for a white noise reference signal.

at about 8 kHz. At this frequency, the DFX divides the adaptation constant by about 4×10^{-5}, which causes the problem of dividing by a small number mentioned earlier. By limiting the smallest number the decorrelation process uses, the algorithm is stabilised and gives better performance than the APFX. This is expected since the APFX equivalently divides by the square root of the power of $\underline{c}$ at each frequency bin while the DFX divides by the whole power. Increasing δ above a certain value scales the adaptation constant lower at other frequencies without performing the required decorrelation at the notch frequency. This may explain the slower initial convergence rate and the smaller final misadjustment of curve 3.

From Fig. 4.30, both the APFX and the DFX in this example decrease the error by 20 dB in about 120000 samples. This corresponds to 2.7 seconds at 44.1 kHz sampling frequency, which is about half the time required for the conventional FX algorithm to reduce the error to the same level. For tracking the listeners' movements using extra white noise (see Section 4.7.1), it is sufficient that the adaptive filter reaches a reasonable solution before the listener starts perceiving the error in the filters due to the movements. Since the adaptive filter in the tracking situation starts off from an initial condition that is close to the optimal solution, and only high frequency coefficients are expected to change, faster convergence time is expected. However, psychoacoustical experiments are necessary to determine whether tracking in 3D sound systems can be performed using adaptive filters.

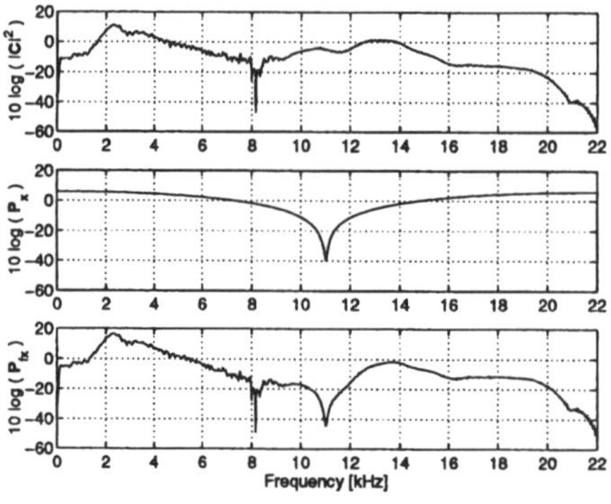

Figure 4.31: The power spectrum of the components $\mathbf{C}$, $\mathbf{X}$ and $\mathbf{X}_f$.

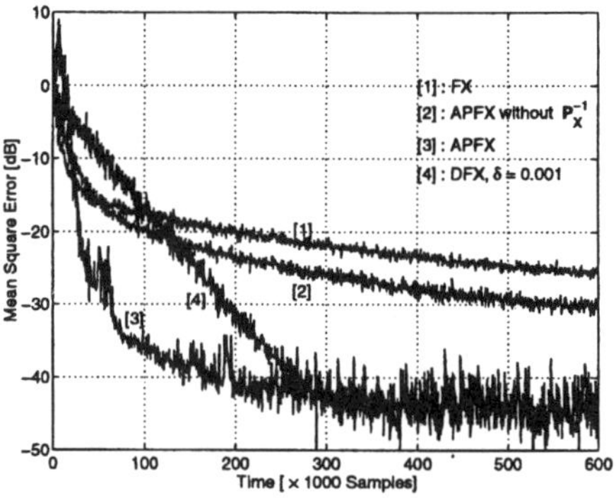

Figure 4.32: The learning curves for an MA(2) reference signal.

4.8.3.2 Performance for a MA(2) Reference Signal

To examine the performance of the different decorrelation methods for colored reference signals, the above mentioned experiment was repeated after filtering the white noise signal by a moving average filter MA(2) with complex conjugate zeros at $\pm\, j\ 0.995$ before using it as a reference signal $\mathbf{X}$. The power spectrum of $\mathbf{C}$, $\mathbf{X}$, and $\mathbf{X}_f$ are shown in Fig. 4.31, which shows the amount of coloration inserted by each component. The adaptive filter $\underline{\mathbf{W}}$ was updated according to each of the above presented algorithms. The learning curves obtained in four cases are shown in Fig. 4.32. Curve 1 is the learning curve for the conventional FX algorithm given by (4.45). Curves 2, 3, and 4 are the learning curves for the APFX update equation (4.53) without multiplying by $\mathbf{P}_X^{-1}$, the APFX update equation (4.53), and the DFX update equation (4.48) with $\mathbf{P}_{X_f} + \delta\, \underline{\mathbf{1}}$, $\delta = 0.001$, respectively.

From Fig. 4.32, the effect of decorrelating each component is clear. Since the notch in the reference signal is spread over more frequencies than that in the secondary path, little improvement is observed by partial decorrelation of $\underline{\mathbf{c}}$ only (curve 2). However, the improvement is evident when the decorrelation of both the reference signal and the secondary path is considered (curve 3). The scaling of the adaptation constant of the fast modes due to adding $\delta = 0.001$ is clear from the difference in the initial convergence rate between curve 3 and 4. Those results are important for tracking of listeners' movements using the control signals as mentioned in Section 4.7.2, where it is important to use a decorrelation method that considers both the reference signal and the secondary path.

4.9 Summary

Several techniques have been introduced in this chapter to deal with moving listeners. Two main approaches have been presented: the first is enlarging the zones of equalisation to cope with small listener movements, and the second is tracking the listener when he moves beyond the equalisation zones.

Enlarging the equalisation zones is achieved by designing the control filters to have a valid solution in wider areas around the control points. Several methods have been presented to reach this goal. These method include approximations to spatial derivative constraints, employing multiresolution control filters, increasing the number of reproduction loudspeakers, and loudspeaker positioning in the listening space. The improvement brought about by each of these methods has been shown by computer simulations in anechoic and reverberant environments.

Derivative constraints require analytical expressions for the acoustic transfer functions, which are difficult to obtain in reverberant environments. Approximations to derivative constraints in reverberant environments have been presented in this chapter. However, these approximations are not valid in severe reverberant conditions. Another approximation to derivative constraints is to control the sound field at discrete points in the proximity of the original required control points (difference constraints). This requires using a microphone at each newly introduced control point, which leads to considerable increase in the computational complexity. Furthermore, the microphones have to be closely positioned to obtain non-splitting zones of equalisation. Although the spatial high-pass filters introduced in this chapter solve the problem of increasing complexity, they still require measuring several transfer functions at adjacent points.

Examining the behaviour of the above mentioned methods shows that the constraints force the control filters to assume solutions that increasingly deviate from the exact solutions with increasing frequency. Similar solutions may be obtained by employing filters having decreasing frequency resolution with increasing frequency. Such filters automatically ignore more spectral details with increasing frequency due to their coarser frequency resolution. The improvement in the robustness of 3D sound systems by using such multiresolution filters is also shown by computer simulations.

The zones of equalisation can also be improved by improving the system controllability and rearranging the acoustic transducers to obtain a well-conditioned system. Using computer simulations, it has been shown that increasing the number of reproduction loudspeakers improves the system robustness in both anechoic and reverberant environments. Positioning the reproduction loudspeakers in the listening space is also studied using simulations.

Tracking the listeners has been approached by updating the control filters to correct the errors introduced due to listeners' movements. Such on-line adaptation requires the adaptive filters to reach their new solutions, which correspond to the current listeners' positions, before the listeners perceive the errors. This requires improving the convergence speed of the conventional adaptive algorithms. Two modifications to the conventional filtered-x algorithm have been introduced for this purpose, the decorrelated filtered-x, and the all-pass filtered-x algorithms. The improvement in convergence speed due to these modifications compared to the conventional algorithm has been shown by computer simulations. On-line estimation of the electro-acoustic transfer functions, that is required for the convergence of the adaptive algorithms, has been also considered in this chapter. Several method for on-line measurements of those transfer functions with and without extra training signals have been discussed.

Chapter 5

Fast Multiresolution Analysis

5.1 Introduction

Many real-time audio applications perform convolution and correlation operations in the frequency domain for efficiency reasons [114, 134]. The FFT is usually used to transform the time signals back and forth to the frequency domain. Interpreting the FFT operation as a filterbank, the filters have constant bandwidth, and their centre frequencies are equally spaced. For audio signals, however, the FFT tends to give too little spectral resolution at low frequencies, and too much spectral resolution at high frequencies, due to the non-uniform spectral resolution of the human auditory system as mentioned in Section 2.1.8. In both models discussed in Section 2.1.8, the human auditory system is considered to perform a constant-Q (percentage bandwidth) analysis in the largest part of the audible frequency range.

Constant bandwidth analysis has several undesired effects in audio signal processing. Since all frequencies are analysed using the same bandwidth, this model is less suitable for representing signals of time-varying spectra [8, 120]. Constant resolution analysis also fails to estimate many of the important psychoacoustical phenomena such as spectral masking. Excessive high frequency resolution can also reveal spectral fine structure that is undetectable by the human ear. Most important for 3D audio systems, constant resolution results in control filters that are sensitive to spatial variations due to the short wavelength at high frequencies, as shown in Chapter 4. Therefore, a better signal representation *resembling* that used by the human auditory system is preferred in audio systems. It is the goal of this chapter to investigate such signal representations. To be of any use in real-time adaptive 3D sound systems, the new signal representation must be calculated as fast as the FFT. Therefore, the study in this chapter is restricted to fast algorithms that can be implemented in real-time.

The specifications a signal representation (transformation) must meet to be considered a candidate in the class of signal representations we are seeking are defined in Section 5.2. In these specifications, the signal transformation is required to support the convolution property, so that filtering operations may be performed efficiently by element-wise multiplication in the transformation domain. This convolution property is explained in more details in Section 5.3.

Several multiresolution signal representations that are of fundamental importance and widely used in audio systems are briefly reviewed in Section 5.4. Most of those methods have been developed in other contexts such as optics, image processing, seismic exploration, and biomedical signal processing. The suitability of those transformations for audio applications in general and real-time adaptive 3D sound systems in particular is also discussed in Section 5.4.

Interpreting the output of an FFT algorithm for B uniformly-spaced samples of a time signal as B uniformly-spaced samples of the signal's spectrum raises the question of what the correct interpretation of the output of an FFT algorithm for B non-uniformly-spaced time samples should be. Intuitively, one tends to say that it must be non-uniformly-spaced samples of the signal's spectrum. If the intuitive answer is correct, this may represent an efficient method for calculating a large class of multiresolution transformations. Using a rigorous mathematical description of unitary axis warping transformations, it is shown in Section 5.5 that the intuitive answer is indeed the correct answer. This axis warping transformations can be implemented simply by non-uniformly sampling the signal under consideration according to a warping function and is, therefore, attractive for real-time applications. Another advantage of this approach is that a family of signal representations is obtained, the characteristics of each member of this family is defined by the warping function. The theoretical aspects of unitary axis warping are discussed in Sections 5.5.1 through 5.5.4. The logarithmic warping function is of special interest for audio applications since it results in a signal representation that has coarser frequency resolution with increasing frequency. In the continuous-time domain, this logarithmic warping maps the Fourier transformation onto the Mellin transformation that proved useful in many fields where scale-invariance is required. Due to this property, the logarithmic warping is studied in detail in Section 5.5.5. Implementation of the warping transformations using non-uniform sampling is presented in Section 5.5.6. Finally, an example of modifying a real-time filtering algorithm to accommodate the warping technique is discussed in Section 5.5.7.

5.2 Specifications

Signal representations (or transformations) relevant to real-time audio signal processing in general, and capable of improving the robustness of 3D sound systems in particular, must meet several requirements. The essential requirements are listed below.

1. **Multiresolution:** The transformation must perform multiresolution spectral analysis similar (but need not be identical) to that performed by the human auditory system. This increases the ability of an audio system to estimate important psychoacoustical phenomena and decreases the system's sensitivity to spatial variations.

2. **Linearity:** Many audio signal processing applications, such as active sound control, rely on the superposition principle, which in turn assumes linearity. Therefore, the signal transformation is required to be linear.

3. **Invertibility:** The transformation must be invertible (non-singular). This is required for real-time signal processing operations such as filtering. Signals are transformed into a spectral domain where filtering is performed by element-wise multiplication and the result is transformed back to the time domain. Without an inverse transformation, no such operation is possible.

4. **Fast implementation:** The discovery of the FFT algorithm made it possible to perform the convolution and correlation operations faster in the frequency domain. Therefore, the chosen transformation is required to keep this speed advantage when those operations are performed in the new transformation domain.

5. **Support for the convolution property:** This is essential if filtering is required to be implemented by element-wise multiplications in the transformation domain. Due to its importance, this property is further explained in Section 5.3.

5.3 The Convolution Property

The *linear convolution* of two sequences $x(n)$ and $h(n)$ each of length N results in the sequence $y_L(n)$ of length $2N-1$ given by

$$y_L(n) = \sum_{i=0}^{2N-1} x(i)\, h(n-i), \qquad n = 0, 1, \cdots, 2N-1. \tag{5.1}$$

On the other hand, the *cyclic convolution* is defined only for periodic signals. Therefore, the cyclic convolution of two sequences $x(n)$ and $h(n)$ each of length N can be calculated only if the two sequences are repeated to construct periodic sequences of period N samples. Alternatively, indexes may be calculated modulo N to obtain the length N sequence $y_C(n)$ of cyclic convolution result,

$$y_C(n) = \sum_{i=0}^{N-1} x((i))_N\, h((n-i))_N, \qquad n = 0, 1, \cdots, N-1, \tag{5.2}$$

where $((\cdot))_N$ represents modulo N arithmetic.

A transformation $\mathbf{T}$ has the *cyclic convolution* property if the transform of the cyclic convolution of two sequences equals the product of their transforms,

$$\underline{\mathbf{Y}}_C = \underline{\mathbf{X}}_C \otimes \underline{\mathbf{H}}_C, \tag{5.3}$$

where $\underline{\mathbf{Y}}_C = \mathbf{T}\,\underline{\mathbf{y}}$, $\underline{\mathbf{X}}_C = \mathbf{T}\,\underline{\mathbf{x}}$, $\underline{\mathbf{H}}_C = \mathbf{T}\,\underline{\mathbf{h}}$, and $\otimes$ denotes element-wise multiplication. This definition assumes that the sequences are periodically extended with period N or the indexes are evaluated modulo N.

For the $N \times N$ transformation matrix $\mathbf{T}$ to satisfy the *cyclic convolution* property, the elements of $\mathbf{T}$ have to satisfy the following three conditions [7]:

1. All $t_{k,m}$ elements of $\mathbf{T}$ must be N^{th} roots of unity, i.e. $(t_{k,m})^N = 1$.

2. All the $t_{k,1}$ elements (the second column of $\mathbf{T}$) must be distinct for the matrix to be non-singular. Since there are only N distinct N^{th} roots of unity, $t_{k,1}$ must be those N distinct roots.

3. The rows of $\mathbf{T}$ are constructed such that $t_{k,m} = (t_{k,1})^m$, i.e. the mth element in any row must be the second element in the same row raised to the mth power.

If $t_{1,1} = \alpha$, $(t_{1,1})^N = 1$ and the rows are arranged such that $t_{k,1} = (t_{1,1})^k$, then the elements of $\mathbf{T}$ can be written as $t_{k,m} = \alpha^{km}, k, m = 0, 1, \cdots, N-1$, and the matrix $\mathbf{T}$ has the following structure:

$$\mathbf{T} = \begin{bmatrix} \alpha^0 & \alpha^0 & \alpha^0 & \cdots & \alpha^0 \\ \alpha^0 & \alpha^1 & \alpha^2 & \cdots & \alpha^{(N-1)} \\ \alpha^0 & \alpha^2 & \alpha^4 & \cdots & \alpha^{2(N-1)} \\ \vdots & \vdots & \vdots & \ddots & \vdots \\ \alpha^0 & \alpha^{N-1} & \alpha^{2(N-1)} & \cdots & \alpha^{(N-1)(N-1)} \end{bmatrix}. \tag{5.4}$$

This structure makes the transformation $\mathbf{T}$ orthogonal and the elements of the inverse transformation $\mathbf{T}^{-1}$ are given by $N^{-1}(t_{k,m})^{-1}$. The structure in (5.4) is the only one supporting the cyclic convolution property. The DFT with $\alpha = e^{-j2\pi/N}$ is the only transformation supporting the cyclic convolution property in the complex number field [7]. Furthermore, any transformation supporting the cyclic convolution property and, therefore, having the structure of (5.4), has a fast computational algorithm similar to the FFT, provided that N is highly composite. Therefore, the DFT is the only transformation in the complex number field that supports the cyclic convolution property and has a fast computational algorithm [7].

It is worthwhile to mention, however, that transformations supporting the cyclic convolution property in other fields do exist. Examples of these transformations are the Fermat Number Transformation (FNT)

[7, 138] and the Mersenne Number Transformation (MNT) [122]. These transformations have similar properties to those known for the DFT and can be used to implement the convolution operation very efficiently [7]. The drawback of these transformations is the lack of a physical meaning coupled to them as the spectral analysis interpretation for the DFT. Since we are not only interested in fast convolution in the transformation domain but also aim at utilising the human auditory system behaviour, which is well understood in the frequency domain, only transformations in the complex number field will be considered here.

The Linear convolution of two finite length sequences can be implemented using cyclic convolution by padding the two sequences with zeros [114]. If support for linear convolution only is desired, a more general class of transformations exists, which does not restrict $t_{k,1}$ to be the Nth roots of unity but still requires that $t_{k,m} = (t_{k,1})^m$. For this class of transformations to be invertible, $t_{k,1}$ must also be distinct. The general structure of the transformation matrix $\mathbf{T}$ of this type is given by

$$\mathbf{T} = \begin{bmatrix} \alpha_0^0 & \alpha_0^1 & \alpha_0^2 & \cdots & \alpha_0^{(N-1)} \\ \alpha_1^0 & \alpha_1^1 & \alpha_1^2 & \cdots & \alpha_1^{(N-1)} \\ \alpha_2^0 & \alpha_2^2 & \alpha_2^4 & \cdots & \alpha_2^{2(N-1)} \\ \vdots & \vdots & \vdots & \ddots & \vdots \\ \alpha_{N-1}^0 & \alpha_{N-1}^{N-1} & \alpha_{N-1}^{2(N-1)} & \cdots & \alpha_{N-1}^{(N-1)^2} \end{bmatrix}. \tag{5.5}$$

The inverse transformation in this case does not have a simple structure and, in general, a fast algorithm for calculating the transformation does not exist. An example of this class of transformations in the complex number field is the non-uniform discrete Fourier transformation discussed in Section 5.4.3.

5.4 Related Work

In this section, some important transformations are briefly mentioned. Starting with transformations that are of fundamental importance, the short time Fourier transformation and the wavelet transformation are discussed in Sections 5.4.1 and 5.4.2, respectively. The non-uniform discrete Fourier transformation is discussed in Section 5.4.3. The digital frequency warping technique has recently found great interest in audio applications, therefore, a detailed discussion of this technique is given in Section 5.4.4. Section 5.4.5 summarises several other published methods to perform non-uniform spectral analysis.

5.4.1 The Short Time Fourier Transformation

A widely used time-frequency representation of signals with time-varying spectra is the Short Time Fourier Transformation (STFT) defined in the continuous-time domain as [8, 119, 120]

$$F(t,f) = \int_{-\infty}^{+\infty} x(\tau)\; h^*(\tau - t)\, e^{-j2\pi f(\tau - t)} d\tau, \tag{5.6}$$

which is the Fourier transform of $x(\tau)$ seen through the window function $h(\tau)$ centred at time t. The STFT can be considered as a similarity measure between the signal $x(\tau)$ and the transformation kernel $h(\tau - t)\, e^{-j2\pi f(\tau - t)}$. The latter is a constant envelope with increasing number of oscillations as the analysis frequency increases. In the frequency domain, this amounts to frequency shifted (modulated) versions of a prototype low-pass filter given by the Fourier transform of the window $h(\tau)$. Therefore, the STFT may be interpreted as a constant bandwidth filterbank with $F(t, f_c)$ being the output of the band-pass filter centred at f_c as a function of time. Provided that the window $h(t)$ is of finite energy, many exact inversion formula exist. Although efficient implementations of the STFT using the FFT algorithm has been previously published [49, 118], the constant bandwidth analysis does not result in a multiresolution spectra that are preferred in audio applications.

5.4.2 The Wavelet Transformation

The Wavelet Transformation (WT) replaces the frequency shifting operation of the STFT by a scaling operation and is, therefore, considered as a time-scale representation rather than time-frequency one. The continuous-time WT is defined as [40]

$$W(t,a) = \frac{1}{\sqrt{|\,a\,|}} \int_{-\infty}^{+\infty} x(\tau)\, h^* \left(\frac{\tau - t}{a}\right) d\tau, \tag{5.7}$$

where the mother wavelet $h(t)$ is localised in time. Similar to the STFT, the WT can be considered as a similarity measure between the signal $x(\tau)$ and the transformation kernel $(1/\sqrt{|a|})\, h\left(\frac{\tau-t}{a}\right)$. The latter is a time-shifted and scaled version of the mother wavelet. The envelope is narrowed as higher frequencies are analysed, while the number of oscillations remains constant. In the frequency domain, this amounts to using a dilated or compressed versions of a band-pass filter whose relative bandwidths are constant (constant-Q filterbank).

Wavelet transformations can be calculated using a fast method, the Fast Discrete Wavelet Transformation (FDWT). The FDWT is performed by successively filtering the signal through a pair of Quadrature Mirror Filters (QMF). This divides the signal into two parts, each covering half the original bandwidth. Both signals are then downsampled by a factor of two. The high frequency part is taken as the highest order wavelet coefficients, while the low frequency part goes through the same process of filtering and downsampling, which yields one more set of wavelet coefficients. The process is repeated until only one sample is left, which is taken as the lowest order coefficient. However, the FDWT produces much coarser resolution at high frequencies than needed in most audio applications. Better resolution may be obtained by decomposing the output of both QMF filters, which is called wavelet packets. This increases the transformation complexity and destroys the constant-Q structure. For this reason, warped wavelets [62, 63] that are developed using the digital frequency warping technique (see Section 5.4.4) have found their applications in audio signal processing.

5.4.3 The Non-Uniform Discrete Fourier Transformation

The Non-uniform Discrete Fourier Transformation (NDFT) [101] may be considered as a generalisation of the DFT. The DFT is obtained by sampling the Z-transform of a sequence $x(n)$ of length N at N equally spaced points on the unit circle in the Z-plane. The NDFT of the same sequence is defined as samples of the Z-transform at N distinct points located arbitrarily in the Z-plane

$$X[z_k] = \sum_{n=0}^{N-1} x(n)\, z_k^{-n}, \qquad k = 0, 1, \cdots, N-1, \tag{5.8}$$

where $X[z_k]$ is the Z-transform of $x(n)$ evaluated at $z = z_k$. The set of points $\{z_k : k = 0, 1, \cdots, N-1\}$ are distinct points arbitrarily chosen in the Z-plane. This can be expressed in a matrix notation as

$$\underline{\mathbf{X}} = \mathbf{D}\,\underline{\mathbf{x}}, \tag{5.9}$$

where

$$\underline{\mathbf{X}} = \begin{bmatrix} X[z_0] \\ X[z_1] \\ \vdots \\ X[z_{N-1}] \end{bmatrix}, \qquad \underline{\mathbf{x}} = \begin{bmatrix} x_0 \\ x_1 \\ \vdots \\ x_{N-1} \end{bmatrix},$$

$$\mathbf{D} = \begin{bmatrix} z_0^0 & z_0^{-1} & z_0^{-2} & \cdots & z_0^{-N+1} \\ z_1^0 & z_1^{-1} & z_1^{-2} & \cdots & z_1^{-N+1} \\ z_2^0 & z_2^{-1} & z_2^{-2} & \cdots & z_2^{-N+1} \\ \vdots & \vdots & \vdots & \ddots & \vdots \\ z_{N-1}^0 & z_{N-1}^{-1} & z_{N-1}^{-2} & \cdots & z_{N-1}^{-N+1} \end{bmatrix}. \tag{5.10}$$

The transformation matrix $\mathbf{D}$ is of the type given in (5.5), therefore, the NDFT supports the linear convolution property only, as mentioned in Section 5.3. This also suggests that the NDFT does not have a fast computation algorithm, which is essential for real-time implementation. Unlike the inverse DFT, which is obtained by calculating the Hermitian transpose of the forward transformation, the inverse NDFT has to be obtained by inverting the matrix $\mathbf{D}$. Since $\mathbf{D}$ is a Vandermonde matrix, the inverse $\mathbf{D}^{-1}$ exists provided that the N sampling points are distinct. Furthermore, if the sampling points are chosen in complex conjugate pairs, then the transformation shows similar symmetry properties as those known for the DFT with odd, even or real sequences. Finally, the NDFT reduces to the DFT when the N sampling points are chosen at equally spaced angles on the unit circle.

Since the NDFT is missing the special symmetry of $e^{\frac{j2\pi k}{N}}$, which is exploited in deriving the different FFT algorithms, it has to be evaluated using matrix multiplication as in (5.9). A straightforward implementation of (5.9) requires N^2 complex multiplications and $N^2 - N$ complex additions. For real signals and when the sampling points z_k are chosen in complex conjugate pairs, only $\frac{1}{2}N+1$ spectral points have to be calculated. This reduces the number of required complex multiplications and

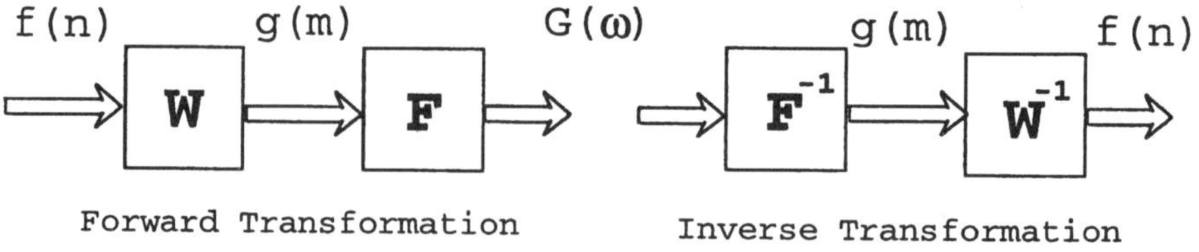

Figure 5.1: Block diagram of the digital frequency warping.

additions to $\frac{1}{2}N^2+N$ and $\frac{1}{2}N^2+\frac{1}{2}N-1$, respectively, which is still proportional to N^2. This high complexity, especially for large N (as in 3D sound systems) makes the NDFT unsuitable for real-time applications. However, the NDFT has found success in off-line applications such as one- and two-dimensional filter design [11, 12].

5.4.4 Digital Frequency Warping

It has been mentioned in Section 5.3 that the DFT is the only transformation in the complex number field that supports the cyclic convolution property and, therefore, the DFT is the only transformation in this field having a fast computational algorithm [7]. To obtain a non-uniform spectral transformation that has a fast computational algorithm, one may search for a linear transformation $\mathbf{W}$ that transforms a uniformly sampled time sequence $f(n)$ to another time sequence $g(m)$ such that the frequency axis Ω defined for $f(n)$ is nonlinearly mapped to the frequency axis ω defined for $g(m)$ [26, 32, 78, 112, 113]. Evaluating the DFT of $g(m)$ results in a uniform spectral sequence $G(\omega)$, which also corresponds to a non-uniform spectral sequence on the Ω axis.

The above mentioned warping principle can be explained from the point of view of the NDFT discussed in Section 5.4.3. The NDFT matrix $\mathbf{D}$ in (5.9) can be factored into the two matrixes $\mathbf{W}$ and $\mathbb{F}$ such that $\mathbf{D} = \mathbb{F}\ \mathbf{W}$, where $\mathbb{F}$ is the DFT matrix and $\mathbf{W}$ is the warping matrix as shown in Fig. 5.1. The warping matrix $\mathbf{W}$ can, therefore, be calculated from $\mathbf{W} = \mathbb{F}^{-1}\ \mathbf{D}$. The required spectral coefficients are obtained by choosing the sampling points $\{z_0, z_1, \cdots, z_{N-1}\}$ in the NDFT on the unit circle. Furthermore, if the symmetry properties of the DFT must be maintained, the sampling points must be chosen in complex conjugate pairs as discussed in Section 5.4.3.

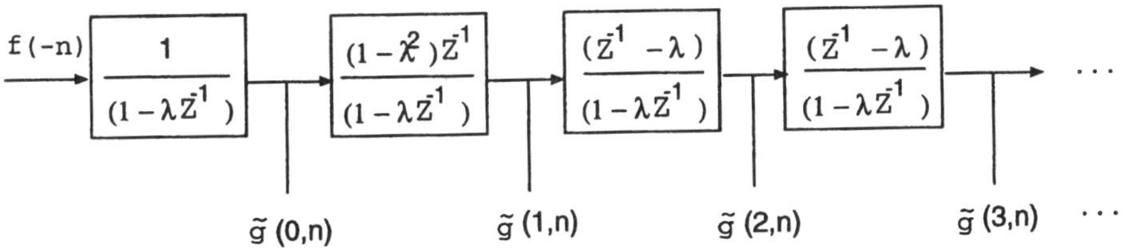

Figure 5.2: A possible implementation of the digital frequency warping transformation **W**.

Since the sampling points in the NDFT are allowed to be arbitrarily chosen, the transformation **W** is, in general, not orthogonal. If the transformation **W** is restricted to be orthogonal, and the mapping is restricted such that ω changes by 2π when Ω changes by 2π, the warping transformation may be implemented by the network shown in Fig. 5.2 [113]. The inverse warping may also be implemented by the same network after replacing each occurrence of λ by $-\lambda$. The sequence $g(m)$ is obtained in Fig. 5.2 by sampling the output of the network at $n = 0$, i.e. $g(m) = \tilde{g}(m, 0)$. For a running evaluation of the spectrum of f(n), the output is sampled every N samples, and the DFT is calculated for the sequence $g(m)$ every N samples. With λ real and between 0 and 1, the warping results in a decreasing frequency resolution with increasing frequency, which resembles the analysis performed by the human auditory system. For λ negative and between -1 and 0, the warping results in an increasing frequency resolution with increasing frequency. For complex values of λ, higher spectral resolution at an intermediate frequency is obtained.

Using the network shown in Fig. 5.2 to generate the warped time sequence $\tilde{g}(m, n)$ requires two real additions and two real multiplications for each $\tilde{g}(m, n)$, except for the first two output points. For $\tilde{g}(0, n)$ only one real addition and one real multiplication are required, while one real addition and two real multiplications are required for $\tilde{g}(1, n)$. If the circuit consists of M sections, the total number of real multiplications and additions per sample is then $2M - 1$ and $2M - 2$, respectively. Since an FFT is calculated every N samples, the total number of real multiplications and additions per FFT is then $2MN - N$ and $2MN - 2N$, respectively, in addition to the operations required for performing the FFT. For $M = N$, the computation complexity is proportional to N^2, which limits the usability of this technique to off-line applications. However, digital frequency warping has found several applications such as modelling HRTFs [84, 85, 86], linear prediction [96], and audio coding [80].

5.4.5 Other Methods

Several authors have published methods to perform non-uniform spectral analysis using block processing algorithms. In 1976, Harris [81] processed the FFT output with spectral windows of constant time duration but adjustable bandwidths centred at the nearest FFT bin to the required analysis frequency. FFT pruning [22, 99, 128, 133] also falls into this category of processing. The pruning is done by discarding samples of the output of the FFT. These methods do not alter the time resolution of the signal and are, therefore, information destructive, which means that they have no inverse transformations.

Another group of authors concentrated on calculating a constant-Q spectral analysis and synthesis. In 1978 Youngberg and Boll [145] presented a constant-Q integral transformation, which is a generalisation of the STFT, by letting the analysis window be a variable in the product of time and frequency. In 1971 and 1972 [74, 75], Gambardella showed that if a signal undergoes a short time spectral analysis via a continuous set of constant-Q band-pass filters, this process can be mathematically represented through an integral transformation that can be inverted by means of the Mellin transformation. The above methods must be implemented as constant-Q filterbanks, which are computationally expensive, making them less attractive for real-time applications. In 1978 and 1983, Kates [88, 89] used an exponentially decaying window whose argument is a constant times the product of time and frequency to calculate the constant-Q integral. For this specific window, the integral can be evaluated using the chirp Z-transformation. Once again, the chirp Z-transformation is too computationally expensive for use in real-time filtering. In 1991 and 1992, Brown [34, 35] has developed a 1/24 octave discrete constant-Q transformation for analysing music signals. This method has no inverse and is, therefore, only suitable for spectral analysis applications.

In 1980, Teaney et al. [135] exploited the perfect fifth numerical coincidence $\frac{3}{2} \cong 2^{\frac{7}{12}}$ to develop what they called the tempered Fourier transformation. They used four analogue-to-digital converters working at sampling rates separated by $2^{\frac{1}{4}} \cong \frac{44}{37}$ to generate a $\frac{1}{12}$ octave analysis. The complexity of using four different sampling rates and the sequence of computations are the difficulties of this method. Complexity and invertibility are not mentioned in [135].

5.5 Unitary Coordinate Transformations

The concept of coordinate transformation, better known as warping, has been used extensively in many applications such as coding, adaptive filters, and dynamic time warping of speech signals. Recently, Baraniuk and Jones [13, 14, 16, 18] have extended the application domain of unitary axis warping to joint signal representations of arbitrary variables including time-frequency and time-scale distributions. The main goal behind coordinate transformation is to represent signals and systems in terms of new basis functions that better suit the application at hand [19, 21, 142]. Audio signal processing is a good example of such applications, since it is known that the human auditory system performs non-uniform spectral analysis on sound waves (see Section 2.1.8). Therefore, audio signal processing systems that interact with the human auditory system should use an internal signal representation similar to that of the human ear. In this section, the fundamentals of unitary warping transformations are introduced. Efficient implementation of warping transformations using non-uniform sampling, and the extension of their applications to real-time filtering are also considered.

Section 5.5.1 introduces the basic principles of unitary operators needed in this section. Duality and unitary equivalence principles, that are introduced in Section 5.5.2 and 5.5.3, respectively, form the tools to deriving the relationships between variables and operators in the time and frequency domains and their warped counterparts. Systems that perform their signal processing in a warped domain are discussed in Section 5.5.4. The multiresolution properties achieved by axis warping transformations are demonstrated for the logarithmic warping in Section 5.5.5. An efficient implementation of the coordinate warping transformations using non-uniform sampling is derived in Section 5.5.6. And finally, an example of tailoring the overlap-save algorithm to accommodate the warping technique is discussed in Section 5.5.7.

In addition to the conventions mentioned in Section 1.5, the following conventions will be used throughout this section. Unless otherwise explicitly stated, all signals are considered to be elements of the Hilbert space of square-integrable functions $L^2(\mathbb{R})$, which has inner product $\langle s,h\rangle = \int_{\mathbb{R}} s(t)h^*(t)dt$ for $s, h \in L^2(\mathbb{R})$ and a second norm $\|h\|^2 = \langle h,h\rangle$. Operators are expressed using boldface capital letters, and the notation $(\mathbf{U}s)(x)$ is used to denote processing the signal s by the operator $\mathbf{U}$ and evaluating the result at x. Throughout this section, the time domain is taken as the default domain of signal representation, therefore, operators are defined in the time domain. The same treatment may be considered for signals and operators defined in the frequency domain with the appropriate interpretation. The symbol $\mathbb{F}$ is used to denote the Fourier operator, which is defined in the continuous-time domain as

$$S(f) = (\mathbb{F}\, s)(f) = \int_{\mathbb{R}} s(t)\, e^{-j2\pi ft}dt. \tag{5.11}$$

5.5.1 Unitary Operators

A unitary operator $\mathbf{U}$ is a linear transformation that maps the Hilbert space onto itself. Unitary operators preserve energy: $\|\mathbf{U}s\|^2 = \|s\|^2$ and inner products: $\langle \mathbf{U}s, \mathbf{U}h\rangle = \langle s,h\rangle$. As a consequence, a unitary operator maps a complete set of orthonormal bases in $L^2(\mathbb{R})$ into another complete set of orthonormal bases in $L^2(\mathbb{R})$. Unitary operators in $L^2(\mathbb{R})$ satisfy the relationship $\mathbf{U}\mathbf{U}^{-1} = \mathbf{U}^{-1}\mathbf{U} = \mathbf{I}$, where $\mathbf{I}$ is the identity operator. In the following development, three classes of unitary operators are needed. The first is the class of unitary axis warping operators defined by variable substitutions. The second is the class of operators associated with physical quantities such as time, frequency and scale variables. Projecting a signal onto the eigenfunctions of the latter class of operators defines the third and last class of unitary operators considered here. A familiar example of this projection is the Fourier transformation which is obtained by projecting a signal onto the eigenfunctions of the time-shift operator. These three classes of unitary operators are examined in more details in this section.

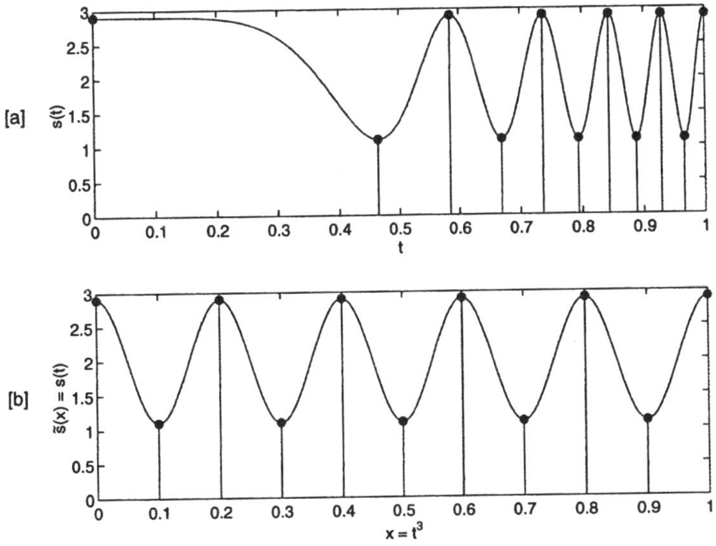

Figure 5.3: [a] A chirp signal $s(t) = 2 + \cos(2\pi 5t^3)$ in the interval $[0, 1]$. [b] The warped version of [a] $\tilde{s}(x) = 2 + \cos(2\pi 5x)$ after applying the variable substitution $x = t^3$.

5.5.1.1 Axis Warping Operators

Given a function $s(t)$ of one independent variable t, performing the variable substitution $t = \gamma(x)$ results in a new function $\tilde{s}(x) = s(t)$ expressed in terms of the variable $x = \gamma^{-1}(t) = \zeta(t)$. The variable x is often referred to as a warped version of t, and the function $\tilde{s}(x)$ as a warped version of $s(t)$. This warping principle is illustrated in Fig. 5.3 for the chirp function $s(t) = 2 + \cos(2\pi Rt^3)$ in the interval $[0, 1]$. Performing the variable substitution $x = \gamma^{-1}(t) = \zeta(t) = t^3$, and expressing the function in terms of the variable x results in $\tilde{s}(x) = 2 + \cos(2\pi Rx)$, which has a constant frequency R as shown in Fig. 5.3-[b]. The axis warping transformed the chirp signal that is widely spread in frequency to a sinusoidal signal that is well localised in frequency. Although this example is simplistic, it suggests the potential use of axis warping in multiresolution analysis as discussed in the next sections.

The variable substitution mentioned above does not preserve the signal energy as can be seen from Fig. 5.3 and is, therefore, not a unitary transformation. A unitary version of the axis warping transformation on

$L^2(\mathbb{R})$ can readily be shown to be [18, 21]

$$\tilde{s}(x) = (\mathbf{U}s)(x) = \mid \gamma'(x) \mid^{1/2} s(\gamma(x)), \tag{5.12}$$

where $\gamma(x)$ is a smooth, monotonic, and one-to-one function, and $\gamma'(x) = d\gamma(x)/dx$. The weighting factor $\mid \gamma'(x) \mid^{1/2}$ preserves the signal energy, making the transformation unitary, and ensures that orthonormal bases in the t-domain are mapped to orthonormal bases in the x-domain.

The effect of axis warping transformations on the signal's spectrum can be seen from the Fourier transform of $\tilde{s}(x)$ with respect to x

$$\tilde{S}(y) = \int_{\zeta(-\infty)}^{\zeta(\infty)} \tilde{s}(x)\, e^{-j2\pi yx}\, dx, \tag{5.13}$$

where y is the dual variable of x (see also Sections 5.5.1.3 and 5.5.2). Note that the limits of the integration cover the interval of support of the variable x. Substituting $x = \gamma^{-1}(t) = \zeta(t)$ in (5.13) and using the relationship $\mid \frac{d\gamma(x)}{dx} \mid = 1/ \mid \frac{d\zeta(t)}{dt} \mid$ we obtain

$$\tilde{S}(y) = \int_{-\infty}^{\infty} s(t) \left[\sqrt{\mid \zeta'(t) \mid}\, e^{-j2\pi y\zeta(t)}\right] dt, \tag{5.14}$$

where $\zeta'(t) = d\zeta(t)/dt$. Equation (5.14) defines a new spectral transformation giving the relationship between the original variable t and the new spectral variable y, opposed to the regular Fourier transformation that connects t with its dual variable f. Both the original Fourier transformation given by (5.11) and the transformation given by (5.14) can be interpreted as measures of similarity between the signal $s(t)$ and the transformation kernel. While the Fourier bases $e^{j2\pi ft}$ are linearly frequency modulated versions of each other, the bases in (5.14) are amplitude modulated (AM) as well as nonlinearly frequency modulated (FM) versions of each other. Frequency modulation defines the analysis values of the new variable y, which is related to the instantaneous frequency at time t $\langle f \rangle_t$ by [45]

$$\langle f \rangle_t = y\,\zeta'(t). \tag{5.15}$$

On the other hand, both types of modulation contribute to the analysis bandwidth of the kernel. The contribution of each modulation type in the analysis bandwidth for a real mapping function $\zeta(t)$ is given by [45]

$$B_{AM}^2 = \frac{1}{4\pi^2}\int\left(\frac{d}{dt}\left[\sqrt{|\zeta'(t)|}\right]\right)^2 dt, \tag{5.16}$$

$$B_{FM}^2 = \int\left(y\zeta'(t)-\langle f\rangle\right)^2 |\zeta'(t)|\, dt, \tag{5.17}$$

where $\langle f\rangle$ is the global average frequency given by

$$\langle f\rangle = y\int \zeta'(t)\ |\zeta'(t)|\, dt. \tag{5.18}$$

From these relationships, it is clear that both the instantaneous frequency and the analysis bandwidth are dependent on the derivative of the mapping function $\zeta'(t)$. Therefore, the key point is the choice of $\zeta(t)$ to achieve the required spectral resolution. Alternatively, starting in the x-domain, and performing the variable substitution $t=\gamma(x)$ results in similar relationships between $\langle y\rangle_x$, f, and $\gamma(x)$.

The relationships between the original time variable t, its dual f and the warped versions x and y are best derived through the duality and unitary equivalence principles introduced in Section 5.5.2 and 5.5.3, respectively. These principles make use of the idea of associating unitary operators to physical variables. Such unitary operators and the expansion onto their eigenfunctions are summarised in Sections 5.5.1.2 and 5.5.1.3, respectively.

5.5.1.2 Operators Associated with Physical Quantities

The idea of associating a variable (such as time or frequency) with an operator is fundamental to the theory of signal representations. Historically, variables have been associated with Hermitian operators [43, 44]. Recently, parameterised unitary operators have also been used [14, 17, 18, 19]. For dual variables, the two representations are equivalent through Stone's theorem [18, 126]. Therefore, only unitary operators correspondence will be considered in the following discussion. In this representation, a variable x is associated with a family of unitary shift operators $\{\mathbf{X}_\xi\}$ parameterised by $\xi\in\mathbb{R}$ and satisfying

$$\mathbf{X}_{\xi_1+\xi_2}=\mathbf{X}_{\xi_1}\mathbf{X}_{\xi_2},\qquad \mathbf{X}_0=\mathbf{I},\qquad \text{and } \mathbf{X}_\xi\to\mathbf{X}_{\xi_1} \text{ as } \xi\to\xi_1, \tag{5.19}$$

Operator	Definition in the time domain	
Time-shift	$(\mathbf{T}_\mu\, s)(t) = s(t-\mu)$,	$s \in L^2(\mathbb{R})$
Frequency-shift	$(\mathbf{F}_\nu\, s)(t) = e^{-j2\pi\nu t}\, s(t)$,	$s \in L^2(\mathbb{R})$
Dilation	$(\mathbf{D}_\sigma\, s)(t) = e^{-\sigma/2}\, s(e^{-\sigma}t)$,	$s \in L^2(\mathbb{R}_+)$

Table 5.1: Time domain definitions of the time-shift, frequency-shift and dilation families of operators.

so that a change in the parameter ξ of the operator $\mathbf{X}_\xi$ directly corresponds to a change in the value of the variable x. Examples of this correspondence are shown in Table 5.1 for the time, frequency and scale variables. The time-shift family of operators $\{\mathbf{T}_\mu : L^2(\mathbb{R}) \mapsto L^2(\mathbb{R})\}$ is associated with the time variable t, the frequency-shift family of operators $\{\mathbf{F}_\nu : L^2(\mathbb{R}) \mapsto L^2(\mathbb{R})\}$ is associated with the frequency variable f, and the scale-shift (dilation) family of operators $\{\mathbf{D}_\sigma : L^2(\mathbb{R}_+) \mapsto L^2(\mathbb{R}_+)\}$, $\mathbb{R}_+ = [0,\infty)$ is associated with the scale variable d [18, 127]. Note that the operators given in Table 5.1 are defined in the time domain and operate on time domain signals. Alternatively, operators may be defined in the frequency domain to operate on frequency domain signals.

5.5.1.3 Eigenfunctions Expansion Operators

The family of unitary operators $\{\mathbf{X}_\xi : \mathcal{H} \mapsto \mathcal{H}\}$, where $\mathcal{H}$ is a closed subset of $L^2(\mathbb{R})$, which corresponds to the variable x possesses the common set of eigenfunctions $\{e_{\mathbf{X}}(y,\cdot) : y \in \mathbb{R}\}$ (which is independent of ξ), and the corresponding set of eigenvalues $\{e^{-j2\pi\xi y}\}$. The expansion of a signal $s(t)$ onto these eigenfunctions defines another unitary transformation $\mathbf{S}_{\mathbf{X}} : \mathcal{H} \mapsto L^2(\mathbb{R})$ given by [21, 127]

$$S(y) = (\mathbf{S}_{\mathbf{X}} s)(y) = \langle s, e_{\mathbf{X}}(y,t)\rangle = \int_{\mathcal{H}} s(t)\; e^*_{\mathbf{X}}(y,t)\; dt, \qquad (5.20)$$

which is the signal representation with respect to the eigenfunctions of $\{\mathbf{X}_\xi\}$. The inverse transformation $\mathbf{S}_{\mathbf{X}}^{-1} : L^2(\mathbb{R}) \mapsto \mathcal{H}$ is given by

$$s(t) = (\mathbf{S}_{\mathbf{X}}^{-1} S)(t) = \langle S, e^*_{\mathbf{X}}(x,t)\rangle = \int_{\mathbb{R}} S(y)\; e_{\mathbf{X}}(y,t)\; dy. \qquad (5.21)$$

The operator $\mathbf{X}_\xi$ is then given by

$$(\mathbf{X}_\xi s)(t) = \int_{\mathbb{R}} e_{\mathbf{X}}(y,t) e^{-j2\pi\xi y} S(y)\, dy. \tag{5.22}$$

Using (5.20) and (5.21), we can write (5.22) more compactly as

$$\mathbf{X}_\xi = \mathbf{S}_{\mathbf{X}}^{-1} \mathbf{\Lambda}_\xi \ \mathbf{S}_{\mathbf{X}}, \tag{5.23}$$

where the diagonal operator $\mathbf{\Lambda}_\xi$ is defined as

$$(\mathbf{\Lambda}_\xi s)(x) = e^{-j2\pi\xi x} s(x), \qquad x \in \mathbb{R}. \tag{5.24}$$

Taking the Fourier transformation of both sides of (5.24) results in the shift property of the Fourier transformation

$$\mathbf{\Lambda}_\xi = \mathbb{F}^{-1} \, \mathbf{\Gamma}_{-\xi} \ \mathbb{F}, \tag{5.25}$$

where $\mathbf{\Gamma}_\xi$ is the shift operator defined as

$$(\mathbf{\Gamma}_\xi s)(x) = s(x - \xi), \qquad x \in \mathbb{R}. \tag{5.26}$$

Alternatively, taking the inverse Fourier transformation of both sides of (5.24) results in the relationship

$$\mathbf{\Lambda}_\xi = \mathbb{F} \, \mathbf{\Gamma}_\xi \ \mathbb{F}^{-1}. \tag{5.27}$$

Therefore, depending on the domain of signal representation, either (5.25) or (5.27) holds. The transformation $\mathbf{S}_{\mathbf{X}}$ given by (5.20) is known to be invariant (up to a phase shift) to the transformation of the signal by the operator $\mathbf{X}_\xi$, for all ξ. That is

$$(\mathbf{S}_{\mathbf{X}} \mathbf{X}_\xi \, s)(y) = (\mathbf{\Lambda}_\xi \mathbf{S}_{\mathbf{X}} s)(y) = e^{-j2\pi\xi y} (\mathbf{S}_{\mathbf{X}} s)(y), \tag{5.28}$$

which can be readily obtained from (5.23) and (5.24). Therefore, the transformation $\mathbf{S}_{\mathbf{X}}$ is referred to as the $\mathbf{X}$-invariant transformation [18, 127], since from (5.28) $|(\mathbf{S}_{\mathbf{X}} \mathbf{X}_\xi s)| = |(\mathbf{S}_{\mathbf{X}} s)|$. The signal representation obtained by $\mathbf{S}_{\mathbf{X}}$ is independent of changes in the signal corresponding to changes in the variable x, in other words, $\mathbf{S}_{\mathbf{X}}$ ignores the transformations of the signals by $\mathbf{X}_\xi$.

The eigenfunctions and the corresponding eigenvalues of the time-shift, frequency-shift and dilation families of operators are given in Table 5.2.

Op.	Eigenfunctions	Eigenvalues
$\{\mathbf{T}_\mu\}$	$\{e_{\mathbf{T}}(f,t) = e^{j2\pi ft} : (f,t) \in \mathbb{R}^2\}$	$\{e^{-j2\pi\mu f} : f \in \mathbb{R}\}_\mu$
$\{\mathbf{F}_\nu\}$	$\{e_{\mathbf{F}}(\tilde{t},t) = \delta(t-\tilde{t}) : (\tilde{t},t) \in \mathbb{R}^2\}$	$\{e^{-j2\pi\nu\tilde{t}} : \tilde{t} \in \mathbb{R}\}_\nu$
$\{\mathbf{D}_\sigma\}$	$\{e_{\mathbf{D}}(c,t) = \frac{e^{j2\pi c \log(t)}}{\sqrt{t}} : (c,t) \in \mathbb{R} \times \mathbb{R}_+\}$	$\{e^{-j2\pi\sigma c} : c \in \mathbb{R}\}_\sigma$

Table 5.2: The eigenfunctions and the corresponding eigenvalues of the time-shift, frequency-shift and dilation families of operators defined in Table 5.1.

From Table 5.2 and the definition of the invariant transformation (5.20), we can recognise $\mathbf{S_T}$ to be the usual Fourier transformation $\mathbb{F}$ defined in (5.11), that is invariant to time-shifts in the sense

$$(\mathbf{S_T}\mathbf{T}_\mu s)(f) = (\mathbb{F}\ \mathbf{T}_\mu s)(f) = e^{j2\pi\mu f}(\mathbb{F}\ s)(f), \tag{5.29}$$

and the transformation $\mathbf{S_F}$ to be the identity operator $\mathbf{I}$ that is invariant to frequency-shifts in the sense

$$(\mathbf{S_F}\mathbf{F}_\nu s)(t) = (\mathbf{I}\mathbf{F}_\nu s)(t) = e^{-j2\pi\nu t} s(t). \tag{5.30}$$

5.5.2 Duality

Given a family of unitary operators $\{\mathbf{X}_\xi\}$ representing the variable x and satisfying (5.19), it was shown in Section 5.5.1.3 that expansion onto the eigenfunctions of $\{\mathbf{X}_\xi\}$ results in the signal transformation $\mathbf{S_X}$ that is $\mathbf{X}$-invariant. It is shown in [127] that there always exists a dual family of operators $\{\mathbf{Y}_\eta\}$ representing the dual variable y, satisfying (5.19) with eigenfunctions expansion transformation $\mathbf{S_Y}$ that is $\mathbf{X}$-covariant in the sense that

$$(\mathbf{S_Y}\mathbf{X}_\xi s)(x) = (\mathbf{S_Y}\ s)(x \pm \xi). \tag{5.31}$$

That is, a transformation of the signal by $\mathbf{X}_\xi$ corresponds to a translation in the signal representation by ξ. The transformation $\mathbf{S_Y}$ measures the contents of x in the signal, and $(\mathbf{S_Y}\ s)(x)$ is the natural signal representation in the x-domain. The $\mathbf{X}$-covariant transformation can readily be derived by substituting (5.25) or (5.27) into (5.23), giving, respectively,

$$\mathbf{X}_\xi = \mathbf{S_X}^{-1}\Lambda_\xi\mathbf{S_X} = \mathbf{S_X}^{-1}\ \mathbb{F}^{\mp 1}\ \mathbf{\Gamma}_{\mp\xi}\ \mathbb{F}^{\pm 1}\ \mathbf{S_X} = \mathbf{S_Y}^{-1}\mathbf{\Gamma}_{\mp\xi}\mathbf{S_Y}, \tag{5.32}$$

where $\mathbf{S_Y}$ is given by

$$\mathbf{S_Y} = \mathbb{F}^{\pm 1}\, \mathbf{S_X}. \tag{5.33}$$

From (5.32) and (5.26), it follows that $\mathbf{S_Y X_\xi} = \mathbf{\Gamma_{\mp\xi} S_Y}$, which is the covariance definition given by (5.31).

The $\mathbf{S_Y}$ transformation is also $\mathbf{Y}$-invariant, that is, it ignores transformations of the signal by the operator $\mathbf{Y}_\eta$. Therefore, $\mathbf{S_Y}$ can be interpreted as a projection onto the eigenfunctions of $\{\mathbf{Y}_\eta\}$ as in (5.20),

$$(\mathbf{S_Y} s)(x) = \langle s, e_{\mathbf{Y}}(x,t)\rangle = \int_{\mathbb{R}} s(t)\; e^*_{\mathbf{Y}}(x,t)\; dt, \tag{5.34}$$

where the set of eigenfunctions $\{e_{\mathbf{Y}}(x,\cdot) : x \in \mathbb{R}\}$ can be obtained from (5.33)

$$e^*_{\mathbf{Y}}(x,t) = \int_{\mathbb{R}} e^*_{\mathbf{X}}(y,t)\; e^{\mp j2\pi xy}\; dy. \tag{5.35}$$

Finally, the family of dual unitary operators $\{\mathbf{Y}_\eta\}$ is given by

$$\mathbf{Y}_\eta = \mathbf{S_Y}^{-1} \mathbf{\Lambda}_\eta \mathbf{S_Y}. \tag{5.36}$$

To illustrate the above duality concept we consider evaluating the dual family of the time-shift operator $\{\mathbf{T}_\mu\}$ defined in Table 5.1. Substituting $e_{\mathbf{X}}(y,t) = e^{j2\pi yt}$ in (5.35), and using the inverse Fourier transformation sign, we obtain $e_{\mathbf{Y}}(x,t) = \delta(t-x)$, which is exactly the eigenfunctions of the frequency-shift operator $\{\mathbf{F}_\nu\}$. From (5.34) we obtain $(\mathbf{S_Y}\, s)(x) = s(x)$, that is, $\mathbf{S_Y} = \mathbf{S_F} = \mathbf{I}$, which satisfies the covariance property (5.31): $(\mathbf{IT}_\mu\, s)(t) = (\mathbf{I}\, s)(t-\mu)$. Therefore, $\mathbf{S_F}$ is $\mathbf{F}$-invariant (see Section 5.5.1.3), $\mathbf{T}$-covariant, and measures the time contents in the signal $s(t)$. The family of unitary operators $\{\mathbf{F}_\nu\}$ is then obtained with the help of (5.36) to be $\mathbf{F}_\nu = \mathbf{I}^{-1}\mathbf{\Lambda}_\nu \mathbf{I} = \mathbf{\Lambda}_\nu = e^{-j2\pi\nu t}$. Similarly, it can be shown that $\mathbf{S_T} = \mathbb{F}$ is $\mathbf{T}$-invariant, $\mathbf{F}$-covariant, and measures the frequency contents in the signal $s(t)$. The above duality principle is summarised in the following Theorem [127]:

Theorem 5.1 *Two families of unitary operators $\{\mathbf{X}_\xi\}$ and $\{\mathbf{Y}_\eta\}$, both satisfying (5.19) and admitting the spectral representation (5.23) are dual if and only if $\mathbf{S_Y}$ is $\mathbf{X}$-covariant and $\mathbf{S_X}$ is $\mathbf{Y}$-covariant. The corresponding variables are also dual. Finally, two complete orthonormal sets $\{e_{\mathbf{X}}(y,\cdot) : y \in \mathbb{R}\}$ and $\{e_{\mathbf{Y}}(x,\cdot) : x \in \mathbb{R}\}$ are dual bases for $\mathcal{H}$ if and only if they are related by (5.35).*

5.5.3 Unitary Equivalence

The shift property of the Fourier transformation (5.25) states that a time signal $s(t)$ can be modulated (frequency shifted) either by multiplying it by $e^{-j2\pi\alpha t}$ or by translating its Fourier transform: $\mathbf{F}_\alpha = \mathbb{F}^{-1}\ \mathbf{T}_{-\alpha}\ \mathbb{F}$. Similarly, (5.27) results in $\mathbf{T}_\beta = \mathbb{F}^{-1}\ \mathbf{F}_\beta\ \mathbb{F}$, which expresses time-shift being equivalent to multiplying $(\mathbb{F}\ s)$ by the phase factor $e^{-j2\pi f\beta}$. This principle of operator equivalence modulo a unitary transformation can be generalised to any unitary operator, which leads to the *unitary equivalence* principle [14, 18, 20].

Definition 5.1 *Two operators* $\widetilde{\mathbf{A}}$ *and* $\mathbf{A}$ *are unitary equivalent if* $\widetilde{\mathbf{A}} = \mathbf{U}^{-1}\mathbf{A}\mathbf{U}$, *with* $\mathbf{U}$ *a unitary transformation.*

In this section, the above unitary equivalence principle is used to establish the relationships between two variables connected by a unitary axis warping transformation of the type (5.12). For this, consider the unitary operator $\mathbf{A} = \mathbf{A}_\alpha$ to be a parameterised unitary operator associated with a physical quantity a (Section 5.5.1.2), and the unitary operator $\mathbf{U}$ a unitary axis warping operator (Section 5.5.1.1). From (5.23), the warped operator $\widetilde{\mathbf{A}}_\alpha$ can be expressed as

$$\widetilde{\mathbf{A}}_\alpha = \mathbf{U}^{-1}\mathbf{A}_\alpha\mathbf{U} = \mathbf{U}^{-1}\mathbf{S}_{\mathbf{A}}^{-1}\Lambda_\alpha\ \mathbf{S}_{\mathbf{A}}\mathbf{U} = \mathbf{S}_{\widetilde{\mathbf{A}}}^{-1}\Lambda_\alpha\ \mathbf{S}_{\widetilde{\mathbf{A}}}, \tag{5.37}$$

where

$$\mathbf{S}_{\widetilde{\mathbf{A}}} = \mathbf{S}_{\mathbf{A}}\mathbf{U} \tag{5.38}$$

is the $\widetilde{\mathbf{A}}$-invariant transformation that is obtained by projecting the signal $s(t)$ onto the eigenfunctions of the warping operator $\widetilde{\mathbf{A}}_\alpha$ (see Section 5.5.1.3). The set of eigenfunctions of $\widetilde{\mathbf{A}}_\alpha$, $\{e_{\widetilde{\mathbf{A}}}(y,\cdot)\}$ can be obtained from the definition of the invariant transformation (5.20)

$$(\mathbf{S}_{\widetilde{\mathbf{A}}}s)(y) = (\mathbf{S}_{\mathbf{A}}\mathbf{U}s)(y) = \langle \mathbf{U}s, e_{\mathbf{A}}(y,\cdot)\rangle = \langle s, \mathbf{U}^{-1}e_{\mathbf{A}}(y,\cdot)\rangle, \tag{5.39}$$

from which

$$e_{\widetilde{\mathbf{A}}}(y,\cdot) = \mathbf{U}^{-1}e_{\mathbf{A}}(y,\cdot). \tag{5.40}$$

Moreover, since $\mathbf{U}$ is a unitary operator, it preserves the inner product in $L^2(\mathbb{R})$ such that $\langle\widetilde{\mathbf{A}},\widetilde{\mathbf{B}}\rangle = \langle\mathbf{A},\mathbf{B}\rangle$ for any two operators $\mathbf{A}$ and $\mathbf{B}$. Therefore, the transformation $\mathbf{U}$ can be interpreted as mapping the family of unitary shift operators $\{\mathbf{A}_\alpha\}$ associated with the variable a onto another family of unitary shift operators $\{\widetilde{\mathbf{A}}_\alpha\}$ associated with the variable $\tilde{a}$. The transformation $\mathbf{S_A}$ that is $\mathbf{A}$-invariant is mapped to $\mathbf{S}_{\widetilde{\mathbf{A}}}$ that can be shown to be $\widetilde{\mathbf{A}}$-invariant. However, the relative angles between operators remain unchanged [18].

Application of the unitary equivalence principle to the time and frequency shift operators $\mathbf{T}_\mu$ and $\mathbf{F}_\nu$ defined in Table 5.1 results in an alternative definition of the duality principle introduced in Section 5.5.2. Noting that the operators $\mathbf{T}_\mu$ and $\mathbf{F}_\nu$ are equivalent to the operators $\mathbf{\Gamma}_\mu$ and $\mathbf{\Lambda}_\nu$, respectively, and using (5.27), the warped time-shift and warped frequency-shift operators $\mathbf{X}_\mu$ and $\mathbf{Y}_\nu$, respectively, can be expressed as

$$\mathbf{X}_\mu = \mathbf{U}^{-1}\mathbf{T}_\mu\mathbf{U} = \mathbf{U}^{-1}\mathbf{\Gamma}_\mu\mathbf{U} = \mathbf{U}^{-1}\ \mathbb{F}^{-1}\ \mathbf{\Lambda}_\mu\ \mathbb{F}\ \mathbf{U} = \mathbf{S}_\mathbf{X}^{-1}\mathbf{\Lambda}_\mu\mathbf{S}_\mathbf{X} \tag{5.41}$$

$$\mathbf{Y}_\nu = \mathbf{U}^{-1}\mathbf{F}_\nu\mathbf{U} = \mathbf{U}^{-1}\mathbf{\Lambda}_\nu\mathbf{U} = \mathbf{S}_\mathbf{Y}^{-1}\mathbf{\Lambda}_\nu\mathbf{S}_\mathbf{Y} \tag{5.42}$$

where

$$\mathbf{S}_\mathbf{X} = \mathbb{F}\ \mathbf{U}, \text{ and } \mathbf{S}_\mathbf{Y} = \mathbf{U}. \tag{5.43}$$

According to (5.33), since $\mathbf{S}_\mathbf{Y} = \mathbb{F}^{-1}\ \mathbf{S}_\mathbf{X}$, then the two warped families of shift operators $\{\mathbf{X}_\mu\}$ and $\{\mathbf{Y}_\nu\}$ are also dual families of unitary operators, the same way $\{\mathbf{T}_\mu\}$ and $\{\mathbf{F}_\nu\}$ are. The results obtained from the duality and unitary equivalence principles when $\mathbf{X}_\mu$ is unitary equivalent to $\mathbf{T}_\mu$ and $\mathbf{Y}_\nu$ is unitary equivalent to $\mathbf{F}_\nu$ are summarised in Fig. 5.4. Note that in this figure, the relationship between the spectral variables f and y are given by an operator $\mathbf{U}$ that is defined in the time domain according to (5.42). Therefore, a signal $S(f)$ is warped to $\tilde{S}(y)$ through the unitary warping transformation $\mathbb{F}\ \mathbf{U}\ \mathbb{F}^{-1}$.

Similarly, using (5.25) in place of (5.27) results in $\mathbf{S}_\mathbf{X} = \mathbb{F}^{-1}\ \mathbf{U}$, $\mathbf{S}_\mathbf{Y} = \mathbf{U}$, and $\mathbf{S}_\mathbf{Y} = \mathbb{F}\ \mathbf{S}_\mathbf{X}$. That is, $\mathbf{X}_\nu$ is unitary equivalent to $\mathbf{F}_\nu$ and $\mathbf{Y}_\mu$ is unitary equivalent to $\mathbf{T}_\mu$. The above result is summarised in the following Theorem,

Theorem 5.2 *Two families of unitary operators $\{\mathbf{X}_\xi\}$ and $\{\mathbf{Y}_\eta\}$, both satisfying (5.19) and admitting the spectral representation (5.23) are dual if and only if they are unitary equivalent to time and frequency operators $\{\mathbf{T}_\mu\}$ and $\{\mathbf{F}_\nu\}$, respectively, via $\mathbf{U} = \mathbf{S}_\mathbf{Y}$ or to $\{\mathbf{F}_\nu\}$ and $\{\mathbf{T}_\mu\}$, respectively, via $\mathbf{U} = \mathbf{S}_\mathbf{X}$.*

Time

quantity	value
Variable	$t \in \mathbb{R}$
Operator	$(\mathbf{T}_\mu\ s)(t) = s(t-\mu)$
Eigenfunc.	$e_{\mathbf{T}}(f,t) = e^{j2\pi ft}$
Eigenval.	$\{e^{-j2\pi f\mu}\}$
Expansion	$\mathbf{S_T} = \mathbb{F}$
Represent.	$\mathbf{S_F} = \mathbf{I}$

Frequency

quantity	value
Variable	$f \in \mathbb{R}$
Operator	$(\mathbf{F}_\nu\ s)(t) = e^{-j2\pi\nu t}\ s(t)$
Eigenfunc.	$e_{\mathbf{F}}(\tilde{t},t) = \delta(t-\tilde{t})$
Eigenval.	$\{e^{-j2\pi\tilde{t}\nu}\}$
Expansion	$\mathbf{S_F} = \mathbf{I}$
Represent.	$\mathbf{S_T} = \mathbb{F}$

Warped Time

quantity	value
Variable	$x \in \mathcal{H}_1$
Operator	$\mathbf{X}_\xi = \mathbf{U}^{-1}\mathbf{T}_\xi\mathbf{U}$
Eigenfunc.	$e_{\mathbf{X}}(y,t) = \mathbf{U}^{-1}e_{\mathbf{T}}$
Eigenval.	$\{e^{-j2\pi y\xi}\}$
Expansion	$\mathbf{S_X} = \mathbb{F}\,\mathbf{U}$
Represent.	$\mathbf{S_F} = \mathbf{U}$

Warped Frequency

quantity	value
Variable	$y \in \mathbb{R}$
Operator	$\mathbf{Y}_\eta = \mathbf{U}^{-1}\mathbf{F}_\eta\mathbf{U}$
Eigenfunc.	$e_{\mathbf{Y}}(x,t) = \mathbf{U}^{-1}e_{\mathbf{F}}$
Eigenval.	$\{e^{-j2\pi\tilde{x}\eta}\}$
Expansion	$\mathbf{S_Y} = \mathbf{U}$
Represent.	$\mathbf{S_X} = \mathbb{F}\,\mathbf{U}$

Figure 5.4: The relationships between the time and the frequency domains and their warped counterparts.

5.5.4 Warped Systems

In this section, the effects of applying a unitary axis warping transformation of the type (5.12) on a Linear Time Covariant (LTC) (known in literature as a linear time invariant) system are discussed. The warping transformation is applied by pre-processing the input signals of an LTC system $\mathbf{P}$ with the unitary transformation $\mathbf{U}$ as shown in Fig. 5.5 [18]. This pre-processing maps the fundamental coordinates of $\mathbf{P}$ (the time and the frequency) to the new concepts x and y defined through the mappings $\mathbf{T} \mapsto \mathbf{X}$, and $\mathbf{F} \mapsto \mathbf{Y}$, as discussed in Section 5.5.3. This is shown by expressing the input and output signals of $\mathbf{P}$ as functions of the warped time variable x instead of the original time variable t (see e.g. Fig. 5.3). The system output $y(x) = (\mathbf{PU}s)(x)$ may be interpreted as the response of the warped-LTC (WLTC) system given by

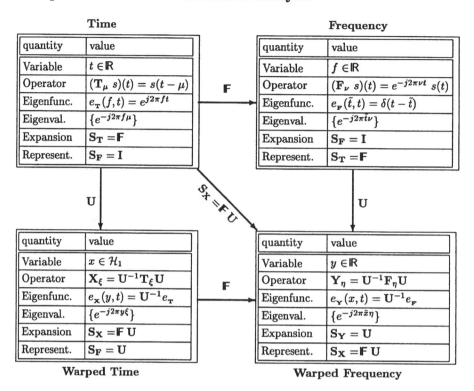

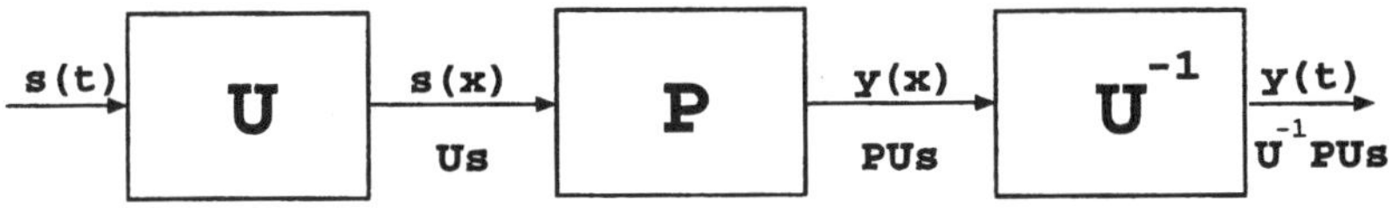

Figure 5.5: A warped linear time covariant system.

PU to the input signal $s(t)$. Since an LTC system is characterised by its covariance to time-shifts, the original system **P** is covariant to **T**: $(\mathbf{PT}_\mu s)(t) = (\mathbf{P}s)(t-\mu)$. However, the WLTC system **PU** is covariant by translation not to **T** but to **X**: $(\mathbf{PUX}_\mu s)(x) = (\mathbf{PU}s)(x-\mu)$. That is, the unitary axis warping transformation **U** maps the **T**-covariant system **P** into the **X**-covariant system **PU**. For instance, the warping function $\gamma(x) = e^x$ in (5.12) maps a linear time covariant system to a linear scale covariant one as will be shortly shown in Section 5.5.5. The power of this approach lies in the following two observations:

- The warped time x is completely defined by the warping operator **U**, therefore, infinitely many internal signal representations inside **P** may be obtained, since infinitely many transformations **U** exist.
- This infinite spectrum of signal representations is realized without requiring a computationally expensive processing unit **P** such as time-varying or time-frequency filters. Well-understood and efficient classical signal processing tools such as the FFT may be used.

These two features make the unitary axis warping transformations attractive for real-time systems, such as adaptive 3D sound systems. The warping technique can be used to implement the multiresolution adaptive filters required for improving the robustness of 3D sound systems as mentioned in Section 4.3. This achieved by pre-processing all input signals of a conventional adaptive 3D sound system by a unitary warping transformation **U**. The warped signals are processed exactly in the same way as before, while in the warped domain. Since the processed signals must be presented to the users as a function of the original time and frequency coordinates, the inverse axis warping $\mathbf{U}^{-1}$ in Fig. 5.5 must be performed. This allows making full use of existing 3D sound systems, while processing the signals in the warped domain maps the time-covariant system to a time-varying one. This enables the system to cope with signals of time-varying spectra, and improves the system ability to predict many of the auditory phenomena that are related to the non-uniform frequency selectivity of the human ears.

5.5.5 Logarithmic Warping and the Mellin Transform

In this section, the mapping function $t = \gamma(d) = e^d$ is considered[1] to illustrate the application and properties of the warping concept discussed in the previous sections. Note that for this mapping function, while $d \in \mathbb{R}$, the variable $t \in \mathbb{R}_+$. Substituting in (5.12), we obtain the unitary warping operator for this warping function,

$$\tilde{s}(d) = (\mathbf{U}s)(d) = e^{d/2}\, s(e^d). \tag{5.44}$$

From the unitary equivalence to the time variable t, the warped eigenfunctions can be obtained from (5.40)

$$e_{\mathrm{D}}(c,t) = \mathbf{U}^{-1}\, e_{\mathrm{T}}(c,t) = \frac{e^{j2\pi c \log(t)}}{\sqrt{t}}. \tag{5.45}$$

The expansion onto these eigenfunctions defines the transformation $\mathbf{S_D} : L^2(\mathbb{R}_+) \mapsto L^2(\mathbb{R})$ as in (5.20)

$$(\mathbf{S_D}\, s)(c) = \langle s, e_{\mathrm{D}}(c,t)\rangle = \int_{\mathbb{R}_+} s(t)\, \frac{e^{-j2\pi c \log(t)}}{\sqrt{t}}\, dt, \tag{5.46}$$

which is the Mellin transform of $s(t)$ [9, 28, 29, 97, 146, 147], that is known to be invariant to dilation as will be soon verified. The shift operator associated with the variable d is then obtained from (5.23)

$$(\mathbf{D}_\sigma\, s)(t) = (\mathbf{S_D^{-1}} \mathbf{\Lambda}_\sigma \mathbf{S_D} s)(t) = e^{-\sigma/2}\, s(e^{-\sigma} t), \qquad s \in L^2(\mathbb{R}_+), \tag{5.47}$$

which is exactly the dilation family of operators $\{\mathbf{D}_\sigma\}$ defined in Table 5.1. Note that $(\mathbf{S_D D}_\sigma\, s)(c) = e^{-j2\pi\sigma c}(\mathbf{S_D}\, s)(c)$, which verifies the Mellin transformation to be **D**-invariant. The dual family of operators is obtained by first deriving the eigenfunctions $\{e_{\mathrm{C}}(d,t)\}$ from (5.35)

$$e^*_{\mathrm{C}}(d,t) = \int_{\mathbb{R}} \frac{e^{-j2\pi c \log(t)}}{e^{d/2}}\, e^{j2\pi c d}\, dc = \frac{\delta(\log(t) - d)}{e^{d/2}}. \tag{5.48}$$

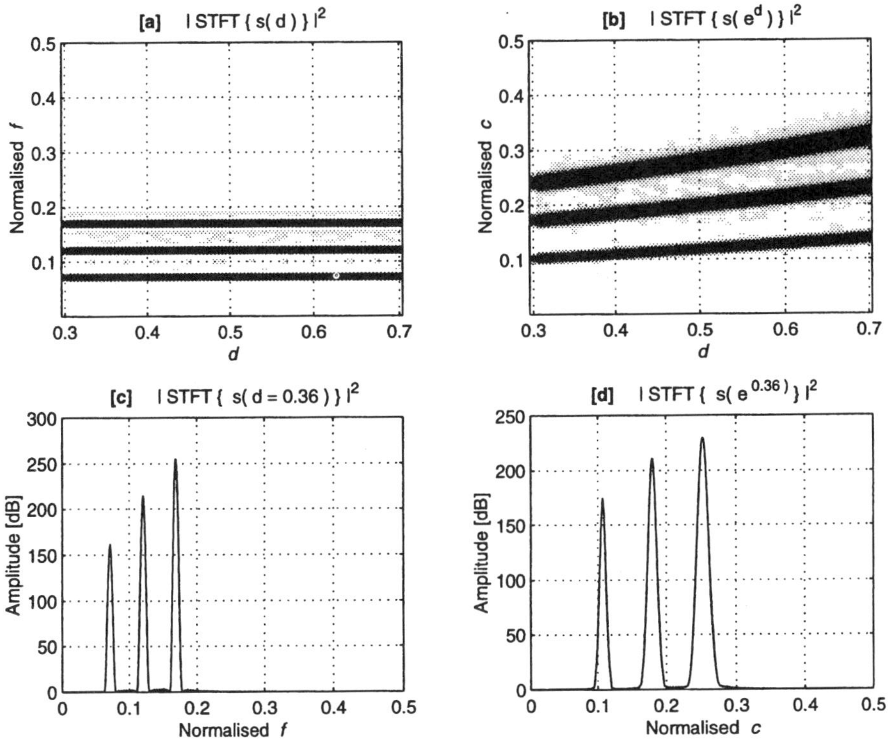

Figure 5.6: [a] The Spectrogram of a signal $s(d)$ composed of three pure tones. [b] The spectrogram of the logarithmically warped version of $s(d)$ for $d_0 = 0.5$ and $\Delta = 0.2$. [c] and [d] are the Fourier transform of the signals in [a] and [b], respectively, evaluated with the window centred at $d = 0.36$.

Using (5.34) and the inverse warping relationship $d = \log(t)$, the transformation $\mathbf{S_C} : L^2(\mathbb{R}_+) \mapsto L^2(\mathbb{R})$, which is the expansion of the signal onto the eigenfunctions $e_{\mathbf{C}}(d, t)$ is obtained

$$(\mathbf{S_C}\, s)(d) = \int_{\mathbb{R}_+} s(t)\, \frac{\delta(\log(t) - d)}{\sqrt{t}}\, dt = e^{d/2} s(e^d), \tag{5.49}$$

which is exactly the unitary warping transformation $\mathbf{U}$ we started off with in (5.44), a result which could be directly obtained from (5.43).

[1]For correct dimensions, a mapping function of the form $t = d_0\, e^{d/d_0}$ should be used, where d_0 is an arbitrary reference value. However, $t = e^d$ will be used here for simplicity.

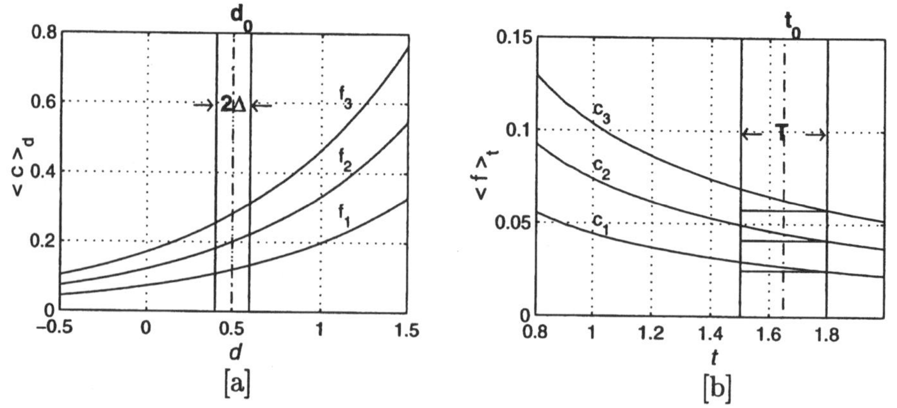

Figure 5.7: [a] The instantaneous Mellin variable $\langle c\rangle_d$ as a function of d in the interval $d = [-0.5, 1.5]$ for the three frequency components of the signal shown in Fig. 5.6-[a]. [b] The instantaneous frequency $\langle f\rangle_t$ as a function of t for $c1 < c2 < c3$.

Note that $\mathbf{S_C}$ is covariant to dilation $(\mathbf{S_C D}_\sigma\, s)(d) = e^{(d-\sigma)/2} s(e^{d-\sigma}) = (\mathbf{S_C}\, s)(d-\sigma)$ and, therefore, measures scale contents in the signal $s(t)$. The dual family of unitary operators $\{\mathbf{C}_\rho\}$ is then given by

$$(\mathbf{C}_\rho s)(t) = (\mathbf{S}_\mathbf{C}^{-1}\boldsymbol{\Lambda}_\rho \mathbf{S_C} s)(t) = e^{-j2\pi\rho\log(t)} s(t), \qquad t \in \mathbb{R}_+, \qquad (5.50)$$

which corresponds to logarithmic modulation [15, 17, 18, 115].

To demonstrate the multiresolution properties of warped signals, we consider applying the unitary warping transformation (5.44) to a simple signal $s(d)$ composed of three pure tones. Starting in the d-domain, the spectrogram (the amplitude squared of the STFT) of this signal in the interval $d = [.3, .7]$ is shown in Fig. 5.6-[a]. Applying the unitary warping transformation (5.44) on $s(d)$ results in the warped signal, the spectrogram of which is shown in Fig. 5.6-[b]. The logarithmic warping mapped the pure tones onto d-varying spectra that have increasing values of c and increasing bandwidth as d increases. A comparison between the Fourier transform of each signal, calculated at a specific d as in Fig. 5.6-[c] and Fig. 5.6-[d], shows that the Fourier transform of the warped signal has increasing analysis bandwidth with increasing c.

The key to understanding the frequency resolution properties of the warped signal in Fig. 5.6-[b] is the instantaneous frequency law given

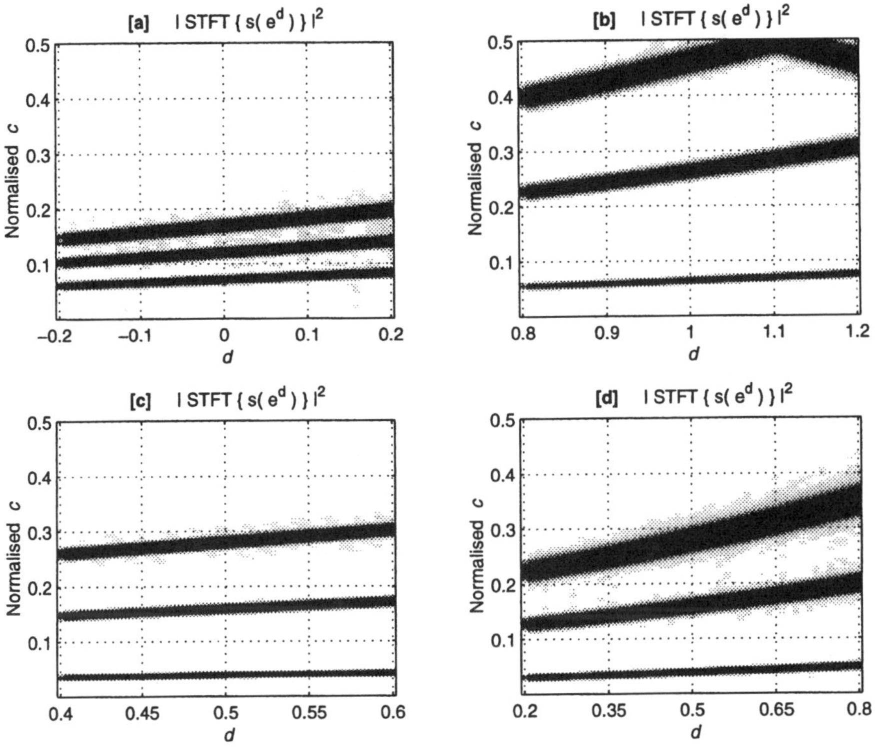

Figure 5.8: The spectrograms of the logarithmically warped version of the signal shown in Fig. 5.6-[a] for $d_0 = 0$ and $\Delta = 0.2$ [a], $d_0 = 1$ and $\Delta = 0.2$ [b], $d_0 = 0.5$ and $\Delta = 0.1$ [c], and $d_0 = 0.5$ and $\Delta = 0.3$ [d].

by (5.15). For the specific warping at hand, the instantaneous Mellin variable $\langle c \rangle_d$ is related to the frequency variable f by

$$\langle c \rangle_d = f\, \gamma'(d) = f\, e^d. \tag{5.51}$$

This relationship is shown in Fig. 5.7-[a] for the frequency components of the signal $s(d)$ shown in Fig. 5.6-[a] in the interval $d = [-0.5, 1.5]$. A comparison between Fig. 5.7-[a] and Fig. 5.6-[b] in the interval $d = [0.3, 0.7]$ suggests that $\langle c \rangle_d$ is the average of the values of the Mellin variable (the warped frequency) that exist in a signal at a specific d, just as the instantaneous frequency $\langle f \rangle_t$ is the average of the frequencies that exist in a signal at a specific time t. For this specific warping, the instantaneous

Mellin variable $\langle c \rangle_d$ as well as its spread (local bandwidth) increase with increasing time. Fortunately, in most practical applications, especially in real-time filtering in a transformation domain, we are interested in processing short segments of signals, which limits this increase. To illustrate, consider that the segment to be analysed is localised at $d = d_0$ and has a duration of 2Δ, which correspond to $t_0 = e^{d_0}$ and an interval $T = e^{d_0+\Delta} - e^{d_0-\Delta}$ in the t-domain, as shown in Fig. 5.7-[b]. In this interval, a given value of c spans a band of frequencies, with higher values of c span more frequencies, which results in wider analysis bandwidths at higher values of the variable c. By controlling the parameters d_0 and Δ, sufficient control on $\langle c \rangle_d$ and its spread can be achieved. The effects of changing the parameters d_0 and Δ on the spectrum of the signal $s(d)$ are shown in Fig. 5.8. For $d_0 = 0$ (Fig. 5.8-[a][2]), the instantaneous Mellin variable coincides with the frequency $\langle c \rangle_d = f$ at $d = 0$. For large values of d_0, the warping may lead to aliasing, as in Fig. 5.8-[b]. This is expected since the frequency component f_3 hits the 0.5 value of the normalised c, as shown in Fig. 5.7-[a] (see also Section 5.5.6).

5.5.6 Axis Warping Using Non-Uniform Sampling

To meet the real-time requirement, an efficient implementation of the unitary warping transformation (5.12) is required to put the warping principle into application. Since DSP systems operate on samples of signals, the starting point is discretising the unitary warping transformation (5.12). Assuming that the sampling process is performed after warping, i.e. in the x-domain, the discrete-time warped signal $\tilde{s}(nT_x)$ (see Fig. 5.3) can be expressed as

$$\tilde{s}(nT_x) = (\mathbf{U}s)(nT_x) = |\, \gamma'(nT_x)\, |^{1/2}\, s(\gamma(nT_x)), \qquad (5.52)$$

where T_x is the sampling period in the x-domain that is considered to comply with the classical sampling theorem, i.e. $T_x \leq 1/(2y_{max})$, and y_{max} is the maximum warped frequency. Implementation of the weighting factor $|\, \gamma'(nT_x)\, |^{1/2}$ in (5.52) forms no difficulties, since $\gamma(x)$ must be chosen a smooth, monotonic, and one-to-one function. In most cases, implementation of this weighting factor can be done off-line, and the result

[2]Different frequencies have been used in this figure to easily show the potential occurrence of aliasing.

stored in a lookup table to be accessed during real-time processing. The second factor of the right hand side of (5.52) can be interpreted as sampling the signal $s(t)$ at sampling moments given by $t_n = \gamma(nT_x)$. That is, the discrete-time warped signal $\tilde{s}(nT_x)$ is a weighted and non-uniformly sampled version of $s(t)$. This can be seen as a generalisation of the discrete Mellin transformation [28, 29], which deals only with exponential sampling. However, such non-uniform sampling may lead to aliasing as in the case encountered in Fig. 5.8-[b], and a method for avoiding this aliasing is required.

Since $\tilde{s}(nT_x)$ is uniformly sampled in the x-domain without aliasing, then the continuous-time signal $\tilde{s}(x)$ can be exactly recovered from the uniformly spaced samples, and so can also the original signal $s(t)$ from its non-uniformly spaced samples. This result is clearly stated in the following Theorem [42]:

Theorem 5.3 *Let $s(t)$ be a band limited function of one variable that is sampled at sampling moments $\{t_n\}$ not necessarily equally spaced. If a one-to-one continuous mapping $x = \zeta(t)$ exists such that $\zeta(t_n) = nT_x$, and if $\tilde{s}(x) = s(\zeta^{-1}(x))$ is band limited to $\omega_o = \pi/T_x$, then the function $s(t)$ can be reconstructed exactly from its samples using the interpolation formula*

$$s(t) = \sum_{n=-\infty}^{\infty} s(t_n) \frac{\sin\left[\frac{\pi}{T_x}\left(\zeta(t) - nT_x\right)\right]}{\left[\frac{\pi}{T_x}\left(\zeta(t) - nT_x\right)\right]}. \tag{5.53}$$

Therefore, for the non-uniform sampling to be a valid implementation of the warping transformation, two conditions have to be satisfied:

1. The variable substitution $x = \zeta(t)$ must be chosen such that the mapping function $\zeta(t)$ is smooth, monotonic and one-to-one and, therefore, invertible with its inverse being $\gamma(x)$.

2. The warped signal is band-limited to y_{max} and the uniform sampling period T_x must be chosen such that $T_x \leq 1/(2y_{max})$. The value of y_{max} for a specific mapping can be calculated from (5.15).

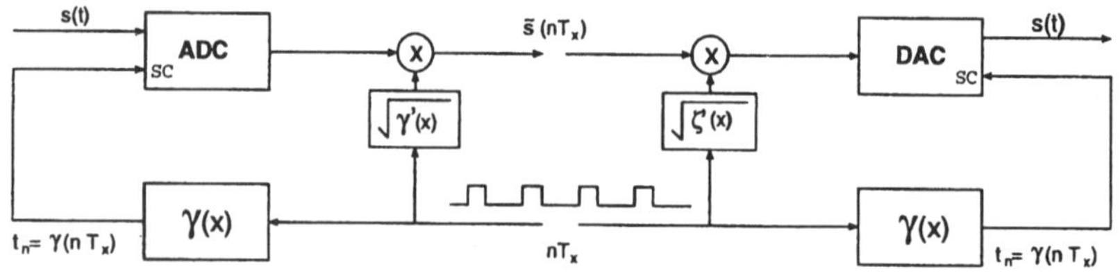

Figure 5.9: Hardware implementation of the warping operation and its inverse warping, sc = start conversion.

The non-uniform sampling suggested by (5.52) may be implemented in software using any of the well known interpolation algorithms. In this case, the transformation complexity depends on the complexity of the interpolation algorithm in use. A more attractive solution for real-time implementation is using a hardware Analogue-to-Digital Converters (ADC) and Digital-to-Analogue Converters (DAC) to perform the warping and inverse warping (reconstruction) operations, respectively. In this case both the ADC and the DAC must be clocked at the non-uniformly spaced sampling moments $t_n = \gamma(nT_x)$ as shown in Fig. 5.9. When the signal to be warped is already in the discrete-time form, an DAC clocked at nT_x is used to convert the signal to continuous-time and then an ADC clocked at $t_n = \gamma(nT_x)$ is used for warping as in Fig. 5.9.

5.5.7 Real-Time Filtering in Warped Domains

As mentioned in Section 5.5.4, a linear time covariant system may be warped by simply warping its input signals as shown in Fig. 5.5. However, since the mapping function is restricted to be smooth and monotonic, the warped time axis must be stretched (or compressed). This leads to compressing (stretching) the frequency axis, and the warped frequencies decrease (increase) without limit, as mentioned in Section 5.5.5. Therefore, warping long signals becomes impractical, and another arrangement has to be found for real-time filtering in warped domains. In this section, the necessary modifications to the overlap-save algorithm to allow real-time filtering in warped domains are discussed.

Real-time frequency domain filtering algorithms successively process short segments of signals. The FFT algorithm is used to transform a block of samples to the frequency domain where convolution is performed by

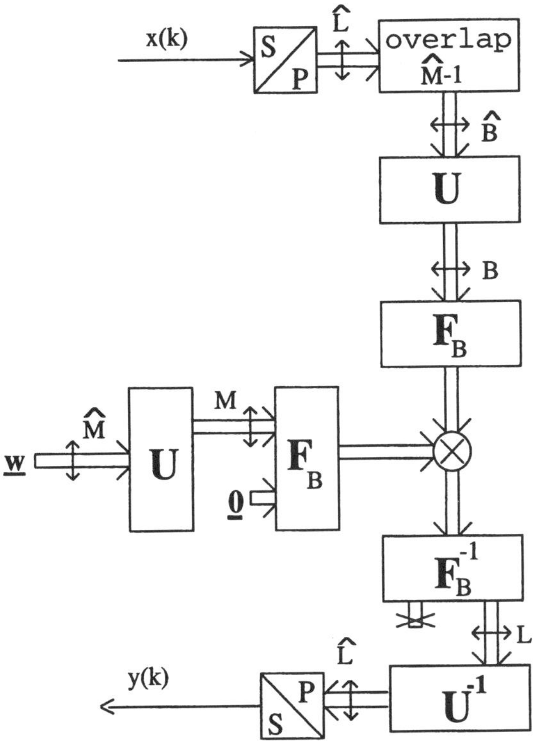

Figure 5.10: Real-time filtering in the warped domain using the overlap-save method.

element-wise multiplication. The multiplication result is then transformed back to the time domain using the inverse FFT. As mentioned in Section 5.5.1.1, calculating the Fourier transformation of a time warped signal amounts to evaluating a new spectral transformation given by $\mathbb{F}$ **U**. Since the FFT considers its input signal to be periodic of a period equal to the block length, only one period need to be warped prior to calculating the Fourier transformation. The inverse warping is then performed on the block of data after calculating the inverse Fourier transformation. This is illustrated in Fig. 5.10 for the overlap-save algorithm [114, 130]. Similar modifications may be applied to other real-time algorithms such as overlap-add and frequency domain adaptive filters.

In Fig. 5.10, the convolution between an infinitely long input signal $x(k)$ and a finite length filter coefficients $\underline{w}$ is calculated in the warped spectral domain. In this Figure, $\hat{M}$ is the filter length, $\hat{L}$ is the number of new input samples per block and $\hat{B} = \hat{M} + \hat{L} - 1$ is the block length in the k-domain. M, L and B are the corresponding variables in the warped domain. The uniformly sampled signal $x(k)$ is stored in an input buffer of length $\hat{B}$ from which $\hat{L}$ samples are new, while $\hat{M} - 1$ samples are taken from the previous block. Every $\hat{L}$ samples, the input buffer is warped using non-uniform sampling as discussed in Section 5.5.6, producing a non-uniformly sampled sequence of length B. This length B vector is transformed to the frequency domain using an FFT algorithm of length B (the $\mathbf{F}_B$ block). The filter coefficients $\underline{\mathbf{w}}$ of length $\hat{M}$ are also warped, padded with zeros, and transformed to the frequency domain. The convolution operation is then performed by element-wise multiplication of the two warped frequency vectors. If a correlation operation is required in place of the convolution, the complex conjugate of one of these two vectors is calculated before performing the multiplication. The product is then transformed back to the time domain, and the first $M - 1$ samples are discarded since they represent a cyclic convolution result. The last L samples are resampled back to the k-domain, giving $\hat{L}$ samples. This vector is finally unbuffered (block P/S), and sent out one sample every clock cycle, giving the linear convolution (or correlation) result $y(k)$.

5.6 Summary

The realisability of real-time multiresolution (adaptive) filters, that are required for improving the robustness of 3D sound systems, has been addressed in this chapter. Since implementation of an (adaptive) filter in a multiresolution transformation domain forces the filter to have the same multiresolution characteristics, the problem reduces to selecting a transformation that possesses the required spectral resolution. The usability of many of the well known and frequently used transformations in real-time adaptive 3D sound systems has been investigated. This study shows that those methods are not suitable for the application in hand due to their high computational complexity, unsuitable frequency resolution, or non-invertibility, which motivated the development of the non-uniform sampling method presented in this chapter.

It has been shown in this chapter that by sampling a signal at non-uniformly spaced time moments and calculating the conventional FFT of those samples, a non-uniformly spaced samples of the spectrum of that signal is obtained. The spectral resolution along the frequency axis of the obtained transform is determined by the non-uniform sampling pattern. This can be seen as an efficient implementation of unitary axis warping transformations and a generalisation of the exponential sampling used to calculate the discrete Mellin transformation. In addition to its efficiency, this method requires minor modifications to convert existing constant resolution systems (such as conventional and adaptive 3D sound systems) to multiresolution ones. This has been demonstrated by modifying the overlap-save algorithm to accommodate the warping technique. The non-uniform spectral resolution of non-uniformly sampled signals has been demonstrated for the logarithmic warping function, and the effects of choosing the centre and duration of the analysis window along the time axis have been presented by examples.

Chapter 6

Conclusions

The objective of the present work is to investigate the application of adaptive methods in creating robust virtual sound images in real-time through loudspeakers. In the previous chapters, the basic principles of adaptive 3D sound systems have been introduced. Efficient implementations, improving robustness, and listeners tracking have been discussed. This chapter summarises the obtained results and highlights the contributions of the present work.

3D sound systems achieve their objectives by filtering monophonic sound signals through a matrix of digital filters. The problem of designing a 3D sound system, therefore, reduces to calculating and implementing those filters. Throughout this book, adaptive filters are used to implement the system filters. In such adaptive 3D sound systems, adaptation of the filters' coefficients is complicated by the matrix of acoustic transfer functions between the reproduction loudspeakers and the listeners' ears (the secondary paths). Due to those transfer functions, the conventional Least Mean Square (LMS) adaptive algorithm becomes unstable and can not be directly used. A modified version, the multiple error filtered-x LMS, must be employed. This algorithm is similar to the LMS except that it correlates the error signals with the input signals filtered by the above mentioned matrix of secondary paths. The disadvantages of the filtered-x algorithm is twofold. First, the system complexity increases rapidly as the number of reproduction loudspeakers, listeners, or audio signals increases. Second, the convergence speed is much degraded due to the coloration of the input signals by the secondary paths.

An efficient alternative to the filtered-x LMS algorithm is the adjoint LMS algorithm. This latter achieves stability by filtering the error signals (instead of the input signals) by the secondary paths, which results in a considerable computational saving for multichannel systems. Further computational saving is gained by implementing the filtering and adaptation operations in the frequency domain. This solves the computational complexity problem of the filtered-x algorithm but does not improve the convergence speed.

To improve the convergence speed of the above mentioned adaptive algorithms, the all-pass filtered-x algorithm has been developed. This algorithm limits the coloration and, therefore, speeds the adaptation process by filtering the input signals (or error signals in the case of the adjoint LMS) through all-pass filters having phase responses equal to those of the secondary paths. This introduces the time delay required for the algorithm stability while keeping the coloration limited without requiring any additional computations.

The matrix of secondary paths not only affects the convergence of the adaptive algorithm, but may also lead to filters that are sensitive to listeners' movements. This is true for adaptive as well as for non-adaptive 3D sound systems. Since the system filters must, partially, implement the inverse of the matrix of secondary paths at all frequencies, this matrix must be non-singular for a solution to exist. If a solution exists, this solution is sensitive to small changes in the matrix if the matrix is ill-conditioned. By examining the condition number of the matrix at each frequency in the frequency range of interest, the optimum positions of the reproduction loudspeakers in a specific listening situation may be investigated. Simulation results in an anechoic environment show that closely spaced stereo loudspeakers result in a matrix that is ill-conditioned at low frequencies and well-conditioned at higher frequencies. As the separation between the two loudspeakers increases, the condition number decreases at low frequencies and spatial aliasing appears at higher frequencies. The decrease in the condition number at low frequencies is transferred equally to the newly introduced peaks due to aliasing. The number of frequencies at which aliasing appears also increases with increasing loudspeaker separation.

Adding a third reproduction loudspeaker to the two-loudspeaker system mentioned above improves the system robustness considerably. However, positioning the third loudspeaker near any of the two original loudspeakers reduces the system to an equivalent two-loudspeaker system.

Similar simulations in reverberant environments show that increasing reverberation increases the system sensitivity to movements. Although similar traces as those obtained in the anechoic simulations could be identified, those traces have higher condition numbers and spread to cover wider frequency bands than in the anechoic case. Adding a third loudspeaker to the stereo system in a reverberant environment improves the system robustness considerably.

From the above mentioned experiments, it may be concluded that in an anechoic environment, closely spaced loudspeakers result in a more robust system at higher frequencies with ill-conditioned filters at low frequencies. On the other hand, widely spaced loudspeakers result in a system that is robust only in narrow frequency bands between the peaks of the condition number resulting from aliasing. The bandwidths of those frequency bands decrease as the separation between the loudspeakers increases. However, no such conclusions can be drawn in reverberant environments. Better system robustness in both anechoic and reverberant environments may be obtained by increasing the number of reproduction loudspeakers.

Since the coefficients of the adaptive filters are dependent on the listeners' positions, the filters become in error when a listener moves. Conventional 3D sound systems use head trackers to correct the introduced errors. In adaptive 3D sound systems, head trackers may not be needed provided that the directional information to be perceived by the listeners is already included in the audio signals so that the system operates as a cross-talk canceller. It is sufficient to update the filters to their new solutions that correct the errors introduced due to movements provided that the adaptive filters converge sufficiently fast. Measurements of the secondary paths that are required by the adaptive algorithm may also be performed on-line (during system operation) using an adaptive identification process. On-line identification and on-line adaptation of the control filters are only possible if microphones are placed inside the listeners' ears.

System robustness to small listeners' movements may be improved by enlarging the zones of equalisation created around the listeners' ears.

Two methods are known in the literature that perform this task. The first is the method of derivative constraints, where the spatial derivatives of the sound field in the vicinity of the listeners' ears are constrained to be zero. This method requires analytic expressions of the acoustic transfer functions to calculate the spatial derivatives. Although approximations of such analytical expressions may be obtained in anechoic environments (with the listeners absent) from the geometry of the listening situation, this becomes more difficult as the reverberation in the listening space increases. In the present work, the method of derivative constraints has been extended by developing approximations to the derivatives which are derived from the measured transfer functions. Although simulations prove the validity of those approximations in moderately reverberant rooms, the approximations fail in severe reverberation situations.

The second method is the method of difference constraints, which controls the sound field at discrete points in the vicinity of each ear to be the same as that required at the eardrum. This is achieved by adding extra microphones at those discrete points and adding a control equation for each microphone. This increases the system complexity considerably and requires more reproduction loudspeakers to better control the sound at the newly introduced microphones. To contain this increase in complexity, the spatial filtering method has been developed. This method reduces the number of control equations by controlling linear combinations of the sound field at discrete points in the vicinity of the listeners' ears. This is on the contrary to explicitly controlling the sound field at each individual point as in the difference constraints method. However, spatial filtering still requires knowledge of the acoustic transfer functions at those individual points to calculate the required linear combinations. This requires inserting microphones at those points and measuring the transfer functions, which is not convenient in many applications. This problem is solved by using multiresolution filters.

Observation of the behaviour of the above mentioned constrained filters shows that the constraints force the filters to assume solutions that are increasingly deviating from the optimum solutions with increasing frequency. This observation is in full agreement with the physics of the problem, since the difference between transfer functions measured at adjacent points in space increases with increasing frequency. An average solution that is valid (but not exact) within small movements (or in a small area in space) may be obtained by ignoring specific spectral details in individual transfer functions. Since specific details increasingly appear with increasing frequency, the average solution may be obtained by employing filters having decreasing frequency resolution with increasing frequency. Such filters automatically ignore more spectral details with increasing frequency due to their coarser frequency resolution. This motivates replacing the conventional constant resolution filters in 3D sound systems by multiresolution ones. Furthermore, the above mentioned physical motivation for using multiresolution filters is also supported by a psychoacoustical motivation. The human auditory system performs a similar non-uniform spectral analysis on sound waves collected by the outer ears. Although ignoring specific details in individual transfer functions has been previously proposed by other authors, to the best of the author's knowledge, multiresolution filters have not been previously suggested.

Non-adaptive multiresolution filters may be calculated by smoothing the acoustic transfer functions using constant-Q (percentage bandwidth) averaging filters prior to using them in calculating the control filters. For adaptive 3D sound systems, the non-uniform frequency resolution may be obtained by updating the filters in a transformation domain that has the required frequency resolution. To maintain the low computational complexity achieved by implementing the filters in the frequency domain, the multiresolution transformation must be calculated as fast as the Fast Fourier Transformation (FFT). An efficient implementation of the unitary warping transformation using non-uniform sampling has been developed for this purpose. By non-uniformly sampling a segment of a signal and calculating the conventional FFT of those samples, non-uniform samples of the spectrum of that segment are obtained. The spectral resolution along the frequency axis of the obtained transform is dependent on the non-uniform sampling pattern. The main advantage of this method is that the architecture of the system is maintained. All that is needed to replace the constant resolution filters by multiresolution ones is to insert a resampling operation before each FFT in the original system. The resampling operation may be implemented in hardware using analogue-to-digital and digital-to-analogue converters.

Bibliography

[1] R.M. Aarts, *On the Design and Psychophysical Assessment of Loudspeaker Systems*, PhD dissertation, Delft University of Technology, 1995, ISBN 90-74445-18-7.

[2] R.M. Aarts, H. He, and P.C.W. Sommen, *Phantom Sources Applied To Stereo-Base Widening Using Long Frequency Domain Adaptive Filters*, Proc. CSSP-96, Mierlo, The Netherlands, 1996, pp. 49-54.

[3] R.M. Aarts, P.C.W. Sommen, A.W.M. Mathijssen, and J. Garas, *Efficient Block Frequency Domain Filtered-x applied to Phantom Sound Source Generation*, AES 104th convention, Amsterdam, May 1998, Preprint No. 4650.

[4] K. Abe, F. Asano, Y. Suzuki, and T. Sone, *A Method for Simulating the HRTF's Considering Head Movement of Listeners*, J. Acoust. Soc. Jpn.(E), Vol. 15, No. 2, 1994, pp. 117-119.

[5] K. Abe, F. Asano, Y. Suzuki, and T. Sone, *Sound Field Reproduction by Controlling the Transfer Function from the Source to Multiple Points in Close Proximity*, IEICE Trans. Fundamentals, Vol. E80-A, No.3, March 1997, pp. 574-581.

[6] AES, *Stereophonic Techniques*, The Audio Engineering Society, New York, 1986.

[7] R.C. Agarwal and C.S. Burrus, *Fast Convolution Using Fermat Number Transforms with Applications to Digital Filtering*, IEEE Trans. on Acoust., Speech, and Sig. Proc., Vol ASSP-22, No. 2, April 1974, pp. 87-97.

[8] J.B. Allen and L.R. Rabiner, *A Unified Approach to Short-Time Fourier Analysis and Synthesis*, Proc. IEEE, Vol. 65, No. 11, Nov. 1977, pp. 1558-1564.

[9] R.A. Altes, *The Fourier-Mellin Transform And Mammalian Hearing*, J. Acoust. Soc. Am., Vol. 63, No. 1, Nov. 1978, pp. 174-183.

[10] F. Asano, Y. Suzuki, and T. Sone *Sound equalization using derivative constraints*, Acoustica, Vol. 82, No. 2, 1996, pp 311-320.

[11] S. Bagchi and S.K. Mitra, *The Nonuniform Discrete Fourier Transform and Its Applications in Filter Design: Part I 1-D*, IEEE Trans. on Circuits and Systems-II: Analog and Digital Signal Processing. Vol 43, No. 6, June 1996, pp. 422-433.

[12] S. Bagchi and S.K. Mitra, *The Nonuniform Discrete Fourier Transform and Its Applications in Filter Design: Part II 2-D*, IEEE Trans. on Circuits and Systems-II: Analog and Digital Signal Processing. Vol 43, No. 6, June 1996, pp. 434-444.

[13] R.G. Baraniuk and D.L. Jones, *New Dimensions in Wavelet Analysis*, Proc. ICASSP-92, San Francisco, March 1992.

[14] R.G. Baraniuk and D.L. Jones, *Warped Wavelet Bases: Unitary Equivalence and Signal Processing*, Proc. ICASSP-93, Minneapolis, March 1993.

[15] R.G. Baraniuk, *A Signal Transform Covariant to Scale Changes*, IEE Electronics Letters, Vol. 29, No. 19, Sept. 17, 1993, pp. 1675-1676.

[16] R.G. Baraniuk and D.L. Jones, *Shear Madness: New Orthonormal Bases and Frames using Chirp Functions*, IEEE Transactions on Signal Processing (Special Issue on Wavelets in Signal Processing), Vol. 41, No. 12, December 1993, pp. 3543-3548.

[17] R.G. Baraniuk, *Beyond Time-Frequency Analysis: Energy Densities in One and Many Dimensions*, Proc. ICASSP-94, Adelaide, Australia, May 1994.

[18] R.G. Baraniuk and D.L. Jones, *Unitary Equivalence: A New Twist on Signal Processing*, IEEE Trans. on Signal. Process., Vol. 43, No. 11, Oct. 1995, pp. 2269-2282.

[19] R.G. Baraniuk, *Warped Perspectives in Time-Frequency Analysis*, IEEE-SP Int. Symposium on Time-Frequency and Time-Scale Analysis, Philadelphia, PA, Oct. 1994.

[20] R.G. Baraniuk, *Covariant Time-Frequency Representations Through Unitary Equivalence*, IEEE Sig. Proc. Letters, Vol. 3, No. 3, March 1996.

[21] R.G. Baraniuk, *Joint Distributions of Arbitrary Variables Made Easy*, Proc. IEEE DSP Workshop, Loen, Norway, Sept. 1996, pp. 394-397.

[22] S. Barash and Y. Ritov, *Logarithmic Pruning of FFT Frequencies*, IEEE Trans. on Sig. Proc., Vol. 41, No. 3, pp. March 1993, 1398-1400.

[23] J. Bauck and D.H. Cooper, *Generalised Transaural Stereo and Applications*, J. Audio Eng. Soc., Vol. 44, No. 9, Sept. 1996, pp. 683-705.

[24] D.R. Begault, *Perceptual Effects of Synthetic Reverberation on Three-Dimensional Audio Systems*, J. Audio Eng. Soc., Vol. 40, 1992, pp. 895-904.

[25] D.R. Begault, *3-D Sound for Virtual Reality and Multimedia*, 1994, ISBN 0-12-084735-3.

[26] D.M. Bell and J.N. Gowdy, *Power Spectral Estimation via Nonlinear Frequency Warping*, IEEE Trans. on Acoust., Speech and Sig. Proc., Vol. ASSP-26, No. 5, Oct. 1978, pp. 436-441.

[27] A.J. Berkhout, D. de Vries, and P. Vogel, *Acoustic Control by Wave Field Synthesis*, J. Acoust. Soc. Am., 1993, No. 5, pp. 2764-2778.

[28] J. Bertrand, P. Bertrand, and J.P. Ovarlez, *Discrete Mellin Transform for Signal Analysis*, Proc. ICASSP-90, Albuquerque, pp. 1603-1606.

[29] J. Bertrand, P. Bertrand, and J.P. Ovarlez, *The Mellin Transform, in The Transforms and Applications Handbook, ed. A.D. Poularikas, Chapter 12*, CRC Press Inc. 1996.

[30] J. Blauert, *Spatial Hearing : The Psychophysics of Human Sound Localisation*, MIT Press, 1997, ISBN 0-262-02413-6.

[31] M.M. Boone, E.N.G Verheijen, and P.F.V. Tol, *Spatial Sound Field Reproduction by Wave Field Synthesis*, J. Audio Eng. Soc., Vol. 43, No. 12, 1995, pp. 1003-1012.

[32] C. Braccini and A.V. Oppenheim, *Unequal Bandwidth Spectral Analysis using Digital Frequency Warping*, IEEE Trans. on Acoust., Speech and Sig. Proc., Vol. ASSP-22, No. 4, Aug. 1974, pp. 236-244.

[33] A.W. Bronkhorst and T. Houtgast, *Auditory Distance Perception in Rooms*, Nature, Vol. 397, Feb. 11, 1999, pp. 517-520.

[34] J.C. Brown, *Calculation of a Constant-Q Spectral Transform*, J. Acoust. Soc. Am., Vol. 89, No. 1, Jan. 1991, pp. 425-434.

[35] J.C. Brown, *An Efficient Algorithm for the Calculation of a Constant-Q Transform*, J. Acoust. Soc. Am., Vol. 92, No. 5, Nov. 1992, pp. 2698-2701.

[36] R.L. Burden and J.D. Faires, *Numerical Analysis*, PWS 1985, ISBN 0-87150-857-5.

[37] G. Chen, M. Abe, and T. Sone, *Evaluation of the Convergence Characteristics of the Filtered-x LMS Algorithm in the Frequency Domain* J. Acoust. Soc. Jpn.(E), Vol. 16, No. 6, 1995, pp. 331-340.

[38] G. Chen and T. Sone, *Effects of Multiple Secondary Paths on Convergence Properties in Active Noise Control Systems with LMS Algorithm*, J. Sound and Vibration, Vol. 195, No. 1, 1996, pp. 217-228.

[39] G. Chen, M. Abe, and T. Sone, *Improvement of the Convergence Characteristics of the ANC System with the LMS Algorithm by Reducing the Effect of Secondary Paths*, J. Acoust. Soc. Jpn. (E), Vol.17, No. 6, 1996, pp. 295-303.

[40] C.K. Chui, *An Introduction to Wavelets*, Academic Press, San Diego, 1992.

[41] G.A. Clark, S.R. Parker, and S.K. Mitra, *A Unified Approach to Time- and Frequency- Domain Realization of FIR Adaptive Digital Filters*, IEEE Trans. on Acoust., Speech and Sig. Proc., Vol. ASSP-31, No. 5, Oct. 1983, pp. 1073-1083.

[42] J. Clark, M. Palmer, and P. Lawrence, *A Transformation Method for The Reconstruction of Functions From Nonuniformly Spaced Samples*, IEEE Trans. Vol. ASSP-33, No. 4, Oct. 1985, pp. 1151-1165.

[43] L. Cohen, *The Scale Representation*, IEEE Trans. on Sig. Proc., Vol. 41, No. 12, Dec. 1993, pp. 3275-3292.

[44] L. Cohen, *Time-Frequency Analysis*, Prentice Hall, 1995.

[45] L. Cohen and C. Lee, *Instantaneous Bandwidth, in Time-Frequency Signal Analysis - Methods and Applications, ed. B. Boashash*, Longman Cheshire, 1992, ISBN 0-470-21821-5.

[46] D.H. Cooper and J. Bauck, *Prospects for Transaural Recording*, J. Audio Eng. Soc., Vol. 37, No. 1/2, Jan./Feb. 1989, pp. 3-19.

[47] D.H. Cooper and J. Bauck, *Head Diffraction Compensated Stereo System*, U.S. Patent 4,893,342, 1990.

[48] L. Cremer and H.A. Müller, *Principles and Applications of Room Acoustics, Volume I*, Applied Science Publications LTD, 1982, English Edition, ISBN 0-85334-113-3.

[49] R.E. Crochiere, *A weighted Overlap-Add Method of Short-Time Fourier Analysis/Synthesis*, IEEE Trans. on Acoustics, Speech and Sig. Proc., Vol. ASSP-28. No. 1, Feb. 1980, pp 99-102.

[50] P. Damaske, *Head-Related Two-Channel Stereophony with Loudspeaker Reproduction*, J. Acoust. Soc. Am., Vol. 50, No. 4, Part 2, 1971, pp. 1109-1115.

[51] W.F. Druyvesteyn and R.M. Aarts, *Personal Sound*, ASA 128 meeting, Paper 3a PP10, 1994.

[52] W.F. Druyvesteyn and J. Garas *Personal Sound*, AES 101st Convention, Los Angeles, Nov. 1996, Preprint 4325.

[53] W.F. Druyvesteyn and J. Garas *Personal Sound*, J. Audio Eng. Soc., Vol. 45, No. 9, Sept. 1997.

[54] R.O. Duda, *3-D Audio for Human Computer Interface*, San Jose State University,
`http://www-engr.sjsu.edu/~knapp/HCIROD3D/3D_home.htm`.

[55] G.P.M. Egelmeers, *Real Time Realization Concepts of Large Adaptive Filters*, PhD dissertation, Eindhoven University of Technology, Nov. 1995, ISBN 90-386-0456-4.

[56] S.J. Elliott, I.M. Stothers, and P.A. Nelson, *A Multiple Error LMS Algorithm and Its Application to the Active Control of Sound and Vibration*, IEEE Trans. on Acoust., Speech and Sig. Proc., Vol. ASSP-35, No. 10, Oct. 1987, pp. 1423-1434.

[57] S.J. Elliott, P. Joseph, A.J. Bullmore, and P.A. Nelson, *Active Cancellation at a Point in a Pure Tone Diffuse Sound Field*, J. Sound and Vibration, Vol 120, No. 1, 1988, pp. 183-189.

[58] S.J. Elliott and P.A. Nelson, *Multiple-Point Equalization in a Room Using Adaptive Digital Filters*, J. Audio Eng. Soc., Vol. 37, No. 11, Nov. 1989, pp. 899-907.

[59] S.J. Elliott, C.C. Boucher, and P.A. Nelson, *The Behavior of a Multiple Channel Active Control System*, IEEE Trans. on Sig. Proc., Vol. 40, No. 5, May 1992, 1041-1052.

[60] L.J. Erikson, *Active Attenuation System with On-Line Modeling of Speaker Error Path and Feedback Path*, U.S. Patent 4,677,676, June 1987.

[61] L.J. Erikson and M.C. Allie, *Use of Random Noise for On-Line Transducer Modeling in an Adaptive Active Attenuation System*, J. Acoust. Soc. Am., Vol. 85, No. 2, Feb. 1989, pp. 797-802.

[62] G. Evangelisa and S. Cavaliere, *The Discrete-Time Frequency Warped Wavelet Transforms*, Proc. ICASSP-97, Munich, pp. 2105-2108.

[63] G. Evangelisa and S. Cavaliere, *Auditory Modeling Via Frequency Warped Wavelet Transform*, Proc. EUSIPCO-98, Rhodes, Sept. 1998, pp. 117-120.

[64] P.L. Feintuch, N.J. Bershad, and A.K. Lo, *A Frequency Domain Model for "Filtered" LMS Algorithms - Stability, Design and Elimination of the Training Mode*, IEEE Trans. on Sig. Proc. Vol. 41, No. 4, May 1993, 1518-1531.

[65] J. Garas, *Personal Sound Using Active Sound Control*, Stan Ackerman Institute Report, Eindhoven University of Technology, The Netherlands, July 1996, ISBN 90-5282-663-3.

[66] J. Garas and P.C.W. Sommen, *Real-Time Convolution and Correlation Using Nonuniform Spectral Analysis and Synthesis*, Proc. IEEE Benelux Sig. Proc. Symposium, Leuven, Belgium, March 1998, pp.95-98.

[67] J. Garas and P.C.W. Sommen, *The All-Pass Filtered-X Algorithm*, Proc. EUSIPCO-98, Rhodes, Sept. 1998.

[68] J. Garas and P.C.W. Sommen, *Improving Virtual Sound Source Robustness Using Multiresolution Spectral Analysis and Synthesis*, AES 105th Convention, San Francisco, Sept. 1998, Preprint No. 4824.

[69] J. Garas and P.C.W. Sommen, *Time/Pitch Scaling Using The Constant-Q Phase Vocoder*, Proc. CSSP-98, Mierlo, The Netherlands, Nov. 1998.

[70] J. Garas and P.C.W. Sommen, *Warped Linear Time Invariant Systems and Their Application In Audio Signal Processing*, Proc. ICASSP-99, Phoenix, March 1999.

[71] J. Garas, *Room Impulse Response, Version 2.1*, `http://www.ses.ele.tue.nl/~garas/room/room.html`.

[72] W.G. Gardner, *3-D Audio Using Loudspeakers*, PhD dissertation, M.I.T. 1997.

[73] W.G. Gardner and K.D. Martin, *HRTF Measurements of a KEMAR*, J. Acoust. Soc. Am., 1997, No.6, pp. 3907-3908.

[74] G. Gambardella, *A Contribution To The Theory Of Short-Time Spectral Analysis With Nonuniform Bandwidth Filters*, IEEE Trans. on Circuit Theory, Vol. CT-18, No. 4, July 1971, pp. 455-460.

[75] G. Gambardella, *The Mellin Transforms and Constant-Q Spectral Analysis*, J. Acoust. Soc. Am., Vol. 66, No. 3, Sept. 1979, pp. 913-915.

[76] R. Genereux, *Adaptive Filters for Loudspeakers and Rooms*, AES 93rd Conventions, Oct. 1992, Preprint 3375.

[77] R. Genereux, *System and Method of Producing Adaptive FIR Digital Filter with Non-Linear Frequency Resolution*, U.S. Patent 5,272,656, Dec. 1993.

[78] J. H. Gilchrist, *The Short-Time Behavior of a Frequency-Warping Power Spectral Estimator*, IEEE Trans. on Acoust., Speech and Sig. Proc., Vol. ASSP-28, No. 2, April 1980, pp. 176-183.

[79] D. Griesinger, *Equalization and Spatial Equalization of Dummy-Head Recordings for Loudspeaker Reproduction*, J. Audio Eng. Soc., Vol. 37, No. 1/2, Jan./Feb. 1989, pp. 20-29.

[80] A. Härmä, *Perceptual Aspects and Warped Techniques In Audio Coding*, Master Thesis, Helsinki University of Technology, May 1997.

[81] F. J. Harris, *High-Resolution Spectral Analysis with Arbitrary Spectral Centers and Arbitrary Spectral Resolutions*, J. Comput. & Elect. Eng. Vol. 3, 1976, pp. 171-191.

[82] W.M. Hartmann, *Listening in a Room and the Precedence Effect, in Binaural and Spatial Hearing in Real and Virtual Environments, ed. R.H. Gilkey and T.R. Anderson, Lawrence*, Erlbaum Associates, 1997, ISBN 0-8058-1654-2.

[83] H. He, *Real Time Generation of Phantom Sound Sources in a Large Frequency Range*, Stan Ackermans Institute Report, Eindhoven University of Technology, The Netherlands, June 1996, ISBN 90-5282-652-8.

[84] J. Huopaniemi and M. Karjalainen, *HRTF Filter Design Based on Auditory Criteria*, Nordic Acoustical Meeting, Helsinki, June 1996.

[85] J. Huopaniemi and M. Karjalainen, *Comparison of Digital Filter Design Methods for 3-D Sound*, IEEE Nordic Sig. Proc. Symposium, Helsinki, Sept. 1996.

[86] J. Huopaniemi, N. Zacharov, and M. Karjalainen, *Objective And Subjective Evaluation Of Head-Related Transfer Function Filter Design*, AES 105th Convention, San Francisco, Sept. 1998, Preprint No. 4805.

[87] P. Joseph, *Active Control of High Frequency Enclosed Sound Fields*, PhD dissertation, University of Southampton, UK, 1990.

[88] J.M. Kates, *Constant-Q Analysis Using the Chirp Z-Transform*, Proc. IEEE Int. Conf. on Acoust., Speech and Sig. Proc., CH-1379, 1978, pp. 314-317.

[89] J.M. Kates, *An Auditory Spectral Analysis Model Using the Chirp Z-Transform*, IEEE Trans. on Acoust., Speech and Sig. Proc., Vol. ASSP-31, No. 1, Feb. 1983, pp. 148-156.

[90] O. Kirkeby, P.A. Nelson, H. Hamada, and F. Orduna-Bustamante, *Fast Convolution of Multichannel Systems Using Regularization*, IEEE Trans. on Speech and Audio Proc., Vol. 6, No. 2, March 1998, pp. 189-194.

[91] O. Kirkeby, P.A. Nelson, and H. Hamada, *The "Stereo Dipole" - A Virtual Source Imaging System Using Two Closely Spaced Loudspeakers*, J. Audio Eng. Soc., Vol. 46, No. 5, May 1998, pp. 899-907.

[92] J. Koring and A. Schmitz, *Simplifying Cancellation of Cross-Talk for Playback of Head-Related Recordings in a Two-Loudspeaker System*, Acoustica, 79, 1993, pp. 221-232.

[93] S.M. Kuo and D.R. Morgan, *Active Noise Control Systems, Algorithms and DSP Implementations*, John Wiley and Sons, 1996, ISBN 0-471-13424-4.

[94] S.M. Kuo and J. Luan, *Multiple-Channel Error Path Modeling with the Inter-Channel Decoupling Algorithm*, Proc. Recent Advances in Active Control of Sound and Vibration, 1993, pp. 767-777.

[95] C. Kyriakakis and T. Holman, *Video-Based Head Tracking for Improvements in Multichannel Loudspeaker Audio*, AES 105th Convention, San Francisco, Sept. 1998.

[96] U. K. Laine, M. Karjalainen, and T. Altosaar, *Warped Linear Prediction In Speech and Audio Processing*, Proc. ICASSP-94, Adelaide, Australia, Vol. III pp. 349-352.

[97] J.S. Lee, M. Burke, J.K. Hammond, *The Theoretical Prediction, Interpretation and Computation of the Fourier-Mellin Transform*

Applied to Sonar Classification of Ships, Proc. ICASSP-90, Albuquerque, pp. 2735-2738.

[98] X. Li and W.K. Jenkins, *Convergence Properties of the Frequency-Domain Block-LMS Adaptive Algorithm*, Proc. ICASSP-95, pp. 1515-1519.

[99] J.D. Markel, *FFT Pruning*, IEEE Trans. on Audio and Electroacoustics, Vol. AU-19, No. 4, Dec. 1971, pp. 305-311.

[100] A.W.M. Mathijssen, *Generation of Phantom Sound Sources with Block Frequency Domain Adaptive Filtering*, Master Thesis, Eindhoven University of Technology, Sept. 1997.

[101] S.K. Mitra, S. Chakrabarti, and E. Abreu, *Nonuniform Discrete Fourier Transform and Its Application in Signal Processing*, Proc. EUSIPCO-92, Brussels, Belgium, Vol. 2, pp. 909-912.

[102] M. Miyoshi and Y. Kaneda, *Inverse Filtering of Room Acoustics*, IEEE Trans. on Acoust., Speech and Sig. Proc., Vol. ASSP-36, No. 2, Feb. 1988, pp. 145-152.

[103] M. Miyoshi and Y. Kaneda, *Active Control of Broadband Random Noise in a Reverberant Three-Dimensional Space*, Noise Control Engineering Journal, Vol. 36, 1991, pp. 85-90.

[104] H. Møller, *Reproduction of Artificial-Head Recordings Through Loudspeakers*, J. Audio Eng. Soc., Vol. 37, No. 1/2, Jan./Feb. 1989, pp. 30-33.

[105] D.R. Morgan, *An Analysis of Multiple Correlation Cancellation Loops with a Filter in the Auxiliary Path*, IEEE Trans. on Acoust., Speech and Sig. Proc., Vol. ASSP-28, No. 4, Aug. 1980, pp. 454-467.

[106] S.S. Narayan, A.M. Peterson, and M.J. Narasimha, *Transform Domain LMS Algorithm*, IEEE Trans. on Acoust., Speech and Sig. Proc., Vol. ASSP-31, No. 3, June 1983, pp. 609-614.

[107] P.A. Nelson, F. Orduna-Bustamante, D. Engler, and H. Hamada, *Inverse Filter Design and Equalization Zones in Multichannel Sound Reproduction*, IEEE Trans. on Speech and Audio Proc., Vol. 3, No. 3, May 1995, pp. 185-192.

[108] P.A. Nelson, H. Hamada, and S.J. Elliott, *Adaptive Inverse Filters for Stereophonic Sound Reproduction*, IEEE Trans. on Sig. Proc. Vol. 40, No. 7, July 1992, 1621-1632.

[109] P.A. Nelson and S.J. Elliott, *Active Control of Sound*, Academic Press Ltd., London, 1992.

[110] P.A. Nelson, F. Orduna-Bustamante, and H. Hamada, *Multichannel Signal Processing Techniques in the Reproduction of Sound*, J. Audio Eng. Soc., Vol. 44, No. 11, Nov. 1996, pp. 973-989.

[111] P.A. Nelson, O. Kirkeby, T. Takeuchi, and H. Hamada, *Sound Fields for the Production of Virtual Acoustic Images*, J. of Sound and Vibration, Vol. 204, No. 2, 1997, pp.386-396.

[112] A.V. Oppenheim, D. Johnson, and K. Steiglitz, *Computation of Spectra with Unequal Resolution Using the Fast Fourier Transform*, Proc. IEEE, Vol. 59, Feb. 1971, pp. 299-301.

[113] A.V. Oppenheim and D.H. Johnson, *Discrete Representation of Signals*, Proc. IEEE, Vol.60, No. 6, June 1972, pp. 681-691.

[114] A.V. Oppenheim and R.W. Schafer, *Discrete-Time Signal Processing*, Prentice-Hall, Englewood Cliffs, 1989, ISBN 0-13-216-771-9.

[115] A. Papandreou, F. Hlawatsch, and G.F. Boudreaux-Bartels The Hyperbolic Class of Quadratic Time-Frequency Representations Part I: constant-Q Warping, the Hyperbolic Paradigm, Properties, and Members, IEEE Trans. on Sig. Proc., Vol. 41, No. 12, Dec. 1993, pp. 3425-3445.

[116] R.D. Patterson and B.C.J. Moore, *Auditory Filters and Excitation Patterns as Representations of Frequency Resolution, in Frequency Selectivity in Hearing, ed. Brian C.J. Moore*, Academic Press, London, 1986, ISBN 0-12-505625-7.

[117] R.D. Patterson, K. Robinson, J. Holdsworth, D. McKeown, C. Zhang, and M.H. Allerhand, *Complex Sounds and Auditory Images, in Auditory Physiology and Perception, ed. Y Cazals, L. Demany and K.Horner*, Pergamon, Oxford, 1992, pp. 429-446.

[118] M.R. Portnoff, *Implementation of the Digital Phase Vocoder Using the Fast Fourier Transform.*, IEEE Trans. on Acoust., Speech and Sig. Proc., Vol. ASSP-24, No. 3, June 1976, pp. 243-248.

[119] M.R. Portnoff, *Time-Frequency Representation of Digital Signals and Systems Based on Short-Time Fourier Analysis*, IEEE Trans. on Acoust., Speech and Sig. Proc. Vol. ASSP-28, No. 1, Feb. 1980, pp. 55-69.

[120] M.R. Portnoff, *Short-Time Fourier Analysis of Sampled Speech*, IEEE Trans. on Acoustics, Speech and Signal Processing, Vol. ASSP-29, No.3, June 1981, pp. 364-373.

[121] V. Pulkki, *Virtual Sound Source Positioning Using Vector Base Amplitude Panning*, J. Audio Eng. Soc., Vol. 45, No. 6, 1997, pp. 944-951.

[122] C. M. Rader, *Discrete Convolution via Mersenne Transforms*, IEEE Trans. on Computers, Dec. 1972, pp. 1269-1273.

[123] K.M. Reichard and D.C. Swanson, *Frequency-Domain Implementation of the Filtered-X Algorithm with On-Line System Identification*, Proc. Conf. on Recent Advances in Active Control of Sound and Vibration, 1993, pp. 562-573.

[124] N. Saito and T. Sone, *Influence of Modeling Error on Noise Reduction Performance of Active Noise Control Systems using Filtered-X LMS Algorithm*, J. Acoust. Soc. Jpn. (E), Vol. 17, No. 4, 1996, pp. 195-202.

[125] N. Saito, T. Sone, T. Ise, and M. Akiho, *Optimal On-Line Modeling of Primary and Secondary Paths in Active Noise Control Systems*, J. Acoust. Soc. Jpn. (E), Vol. 17, No. 6, 1996, pp. 275-283.

[126] A.M. Sayeed and D.L. Jones, *Equivalence of Generalised Joint Signal Representations of Arbitrary Variables*, IEEE Trans. on Sig. Proc., Vol. 44, No. 12, Dec. 1996, pp. 2959-2970.

[127] A.M. Sayeed and D.L. Jones, *Integral Transforms Covariant to Unitary Operators and their Implications for Joint Signal Representations*, IEEE Trans. on Sig. Proc., Vol. 44, No. 6, June 1996, pp. 1365-1377.

[128] D.P. Skinner, *Pruning the Decimation In-Time Algorithm*, IEEE Trans. on Acoust., Speech and Sig. Proc., April 1976, pp. 193-194.

[129] S.D. Snyder and C.H. Hansen, *The Influence of Transducer Transfer Functions and Acoustic Time Delays on the Implementation of the LMS Algorithm in Active Noise Control Systems*, J. of Sound and Vibration, Vol. 141, No. 3, 1990, pp. 409-424.

[130] P.C.W. Sommen, *Adaptive Filtering Methods*, PhD dissertation, Eindhoven University of Technology, The Netherlands, June 1992, ISBN 90-9005143-0.

[131] P.C.W. Sommen, R.M. Aarts, A.W.M. Mathijssen, J. Garas, and H. He *Efficient Frequency Domain Filtered-x Realization of Phantom Sources*, Proc. CCSP-97, Mierlo, The Netherlands.

[132] P.C.W. Sommen and J. Garas, *Using Phase Information to Decorrelate the Filtered-X Algorithm*, Proc. ICASSP-98, Seattle, pp. 1397-1400.

[133] H.V. Sorensen and C.S. Burrus, *Efficient Computation of the DFT with only a Subset of Input or Output Points*, IEEE Trans. on Sig. Proc., March 1993, pp. 1184-1200.

[134] T.G. Stockham, *High-Speed Convolution and Correlation*, Proc. Spring Joint Computer Conf., 1966, pp 229-233.

[135] D.T. Teaney, V.L. Moruzzi, and F.C. Mintzer, *The Tempered Fourier transform*, J. Acoust. Soc. Am., Vol. 67, No. 6, June 1980, pp. 2063-2067.

[136] W.R. Thurlow and P.S. Runge, *Effects of Induced Head Movements on Localisation of Direct Sound*, J. Acoust. Soc. of Am., Vol. 42, 1967, pp. 480-487.

[137] W.R. Thurlow, J.W. Mangels, and P.S. Runge, *Head Movements During Sound Localisation*, J. Acoust. Soc. of Am., Vol. 42, 1967, pp. 489-493.

[138] T.K. Truoong, I.S. Reed, C.S. Yeh, and H.M. Shao, *A Parallel Architecture for Digital Filtering Using Fermat Number Transforms*, IEEE Trans. on Computers, Vol. C-32, No. 9, Sept. 1983, pp. 874-877.

[139] S. Uto, H. Hamada, T. Miura, P.A. Nelson, and S.J. Elliot, *Audio Equalization using Multi-Channel Adaptive Digital Filters*, Int. Symp. on Active Control of Sound and Vibration, Tokyo, April 1991, pp. 421-426.

[140] A.C.P. van Meer, J. Garas, P.C.W. Sommen, and J.L. van der Laar, *Implementation of a Broad Band Multi Point Acoustic Noise Canceller*, Proc. ICSPAT-98, Toronto, pp. 1-5.

[141] E.A. Wan, *Adjoint LMS: An Efficient Alternative to the Filtered-X LMS and Multiple Error LMS Algorithm*, Proc. ICASSP-96, pp. 1842-1845.

[142] A. Wang, *Instantaneous and Frequency-Warped Signal Processing Techniques for Auditory Source Separation*, PhD dissertation, Stanford University, 1994.

[143] F.L. Wightman and D.J. Kistler, *Headphone Simulation of Free-Field Listening, I : Stimulus Synthesis*, J. Acoust. Soc. of Am., Vol. 85, No. 2, Feb. 1989, pp. 858-867.

[144] B. Widrow and S.D. Stearns, *Adaptive Signal Processing*, Prentice-Hall Inc., Englewood Cliffs, 1985, ISBN 0-13-004029-0

[145] J.E. Youngberg and S.F. Boll, *Constant-Q Signal Analysis and Synthesis*, Proc. ICASSP-78, pp. 375-378.

[146] K.P. Zimmermann *Mellin Transforms of Closed Curves and One Dimensional Functions: Direct Computation and Scaling Properties*, Proc. IEEE Vol. 75, No 6, June 1987, pp. 859-860.

[147] P.E. Zwicke and I. Kiss, *A New Implementation of the Mellin Transform and Its Application to Radar Classification of Ships*, IEEE PAMI Vol. 5, No. 3, March 1983, pp 191-199.

[148] E. Zwicker and H. Fastl, *Psychoacoustics: Facts and Models* , Springer-Verlag, 1990, ISBN 3-540-52600-5.

Index

R

S

T